Biological Invasions in the Hudson River Basin

Biological Invasions in the Hudson River Basin: An Inventory and Historical Analysis

Edward L. Mills and Mark D. Scheuerell
Department of Natural Resources
Cornell University Biological Field Station
900 Shackelton Point Road
Bridgeport, New York 13030

James T. Carlton
Maritime Studies Program
Williams College-Mystic Seaport
Mystic, Connecticut 06355

David L. Strayer
Institute of Ecosystem Studies
Millbrook, New York 12545

New York State Museum Circular No. 57

The University of the State of New York
The State Education Department

Printed in the United States of America

Copies may be ordered from:
Publication Sales
New York State Museum
Albany, New York 12230
Phone: (518) 449-1404

Library of Congress Catalog Card Number: 97-61576

ISSN: 1052-2018
ISBN: 1-55557-207-3

Book design: Kathy Daigler
This book is printed on acid-free paper.

Contents

Contents

List of Figures

List of Tables

ACKNOWLEDGMENTS

The authors would like to thank the following people who made various contributions to this working inventory of exotic species in the Hudson River basin: D. Bath, J. Baxter, J. Chrisman, R. Daniels, V. Dawson, C. deQuillfeldt, R. Emerson, J. Enck, P. Geoghegan, C. George, J. Jenkins, F. Keane, C. Keene, W. Keller, M. Keser, E. Kiviat, C. Letts, H. Marshall, N. McBride, R. Mitchell, W. Nieder, M. Pace, B. Peckarsky, C. Reimer, R. Schmidt, C. Secor, C. Sheviak, C. Smith, D. Smith, J. Spitsbergen, G. Stevens, E. Stoermer, R. Stuckey, and B. Swift. We thank Eric Kiviat for his thorough review. Cover illustration by Patricia Kernan. Design by Kathy Daigler. We also thank J. Waldman and D. Suszkowski and the Hudson River Foundation for funding (Grant 027/92A) this work. Contribution number 177 of the Cornell Biological Field Station.

INTRODUCTION

Exotic species constitute one of the most serious of all threats to indigenous biota (Drake et al. 1989; Elton 1958; Hengeveld 1989; Kornberg and Williamson 1986; Mooney and Drake 1986; Pieterse and Murphy 1990). Portions of the earth's biota are rapidly becoming homogenized due to the introduction of species outside their natural range (Daniels 1993). Exotic plant species typically characterize 10-30% of the flora of most regions, and within these regions, exotic plants may comprise 90% or more of the plant biomass (Heywood 1989). The effects of these biological invasions on native communities can be dramatic and wide-ranging and include not only changes in species composition, biomass, and distributions, but also changes in nutrient cycling and energy transfer pathways and rates. Significant indirect effects of species invasions can be transmitted through communities and ecosystems to have an impact on biota at different trophic levels (Mills et al. 1987). Furthermore, exotic species may have extensive impacts on economic and industrial productivity and human health.

Nearly four centuries of European colonization and commercial development in northeastern North America have set the stage for fundamental biological alterations and modifications of the Hudson River system. These modifications largely fall into two categories: (1) the destruction and eradication of natural habitats and native flora and fauna through urbanization, industrialization, and exploitation, and (2) the introduction of non-native species. These modifications are linked; anthropogenic environmental alterations enhance invasion success (Drake et al. 1989; Mooney and Drake 1986). In the Great Lakes, for example, human-mediated environmental activities have led to the introduction of more than 139 species of exotic animals and plants, many of which now dominate local communities (Mills et al. 1993). Increasing industrialization and commercialization have indirectly led to increased rates of invasions, such that over one-third of the non-indigenous species in the Great Lakes have been discovered in the past 30 years. This trend coincides with the opening of the St. Lawrence Seaway in 1959 (Mills et al. 1993).

Our inventory of exotic species focuses on the Hudson River drainage, a large North American river ecosystem draining into the Atlantic Ocean. For the Hudson River itself, our inventory of exotic species includes only the tidal freshwater portion of the river north of river mile 30 (48 km) (Figure 1). Hydrographically and topographically the river flows 507 km from its source, Lake Tear of the Clouds in the Adirondacks, to Manhattan. The entire Hudson

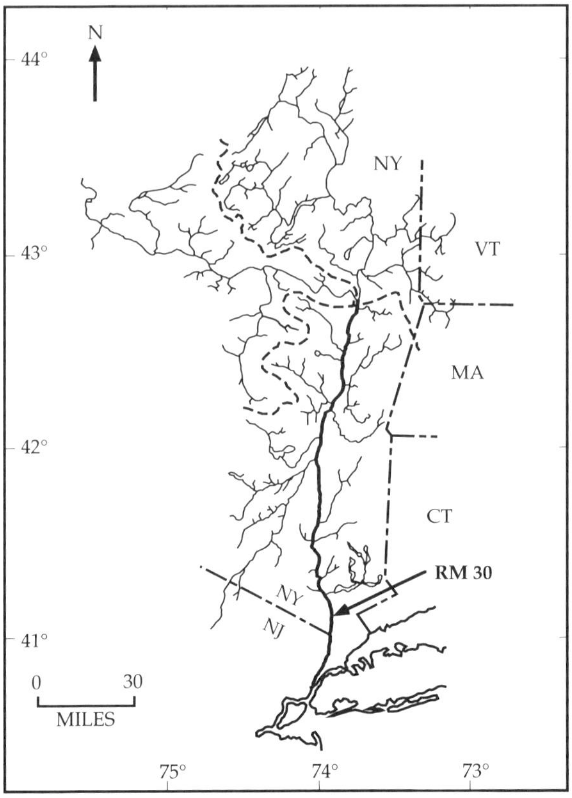

Figure 1. Map of the Hudson River drainage basin.

River drainage basin is large (34,615 km^2) and drains parts of five states (New York, New Jersey, Massachusetts, Connecticut, and Vermont) as well as six physiographic regions (the Canadian Shield, the Folded Appalachians, the Catskills, the Hudson Highlands, the New England Upland, and the New Jersey Lowland) (Kammen 1975).

The purpose of this study is to present a comprehensive inventory of the introduced flora and fauna of the Hudson River drainage basin (Figure 1). This list includes fishes, invertebrates, and aquatic plants that have entered the Hudson River basin since the early 1800s. Our inventory is similar to that described for the Laurentian Great Lakes (Mills et al. 1993) and provides for direct comparison with the Hudson River basin. We chose to focus on published literature, and we therefore did not do a rigorous review of herbarium and museum specimens. We do not include terrestrial plants such as multiflora rose (*Rosa multiflora*) and honeysuckle (*Lonicera x bella*), even though they are non-indigenous to the Hudson River basin. We exclude the marine invaders that occur south of river mile 30 (48 km) (e.g., the European green crab, *Carcium maenas*; the Japanese orange-striped sea anemone, *Haliplanella lineata*; the European periwinkle, *Littorina littorea*; and the southern United States barnacle, *Balanus subalbidus*). We also exclude vertebrate exotics such as the mute swan, the Canada goose, and the European starling. Although exotic species have been successfully invading the Hudson River for several centuries, there has been no previous attempt to review and analyze the history of biological invasions in the region. Previous studies have identified exotic species within certain taxonomic groups (e.g., Countryman 1970; Kiviat 1987; Smith 1985; Smith and Lake 1990; Strayer 1987), but no comprehensive list of aquatic non-indigenous species has been compiled. Here we present such an inventory for the Hudson River drainage basin.

Mechanisms of Introduction of Exotic Species into the Hudson River Basin

The Hudson River basin offers a classic example of a New World ecosystem subjected to four centuries of European exploration, colonization, intensive urbanization, and the associated activities that have led to the accidental or intentional release of exotic species from around the world. From the earliest colonial visits in the seventeenth century, which brought the seeds of marsh plants in livestock feed and fouling hydroids on carrack hulls (Carlton 1992c; Carlton and Hodder 1995) to the late twentieth century release of ballast water from huge ocean-going bulk carriers (Carlton 1985; Carlton and Geller 1993), the Hudson has been a site for the continuous release and successful establishment of non-indigenous aquatic plants and animals.

Early Viking explorations aside, the Hudson River system was potentially first regularly exposed to the invasion of European organisms in the late fifteenth and early sixteenth centuries, when explorers such as Cabot (in 1498) and Gomes (in 1525) visited the river's mouth, and when Verrazano sailed far enough inland in 1524 to see the outline of the Palisades (Bruce 1907; Kammen 1975). On the bottoms of their sailing vessels were presumably European fouling organisms, and in their holds, potentially discardable into the river, were dry ballast, dunnage, and spoiled foodstuffs. Exploration and colonization increased dramatically in the seventeenth century. Henry Hudson sailed all the way to present-day Albany in 1609, and numerous European vessels visited the river system in the decades to follow. For many years, the river was the major transport corridor. In 1609, the river was described as follows: "The road is very neere [to the banks], and very good for all winds, save an east north-east wind" (Juet 1609 in Kammen 1975:3). Consequently, as Kammen (1975:3) noted, "For generations to come, 'the road' would refer to the ships' channel, for these were waterborne communities." There is little doubt that a number of the common marsh weeds we list in this work entered the New York region during these first few centuries of commerce. However, no scientific descriptions of the botany or zoology of the river system became available until two centuries later; thus thorough knowledge of all these early presumed invasions is lost.

For the period from 1600 to 1800, thousands of vessels visited New York City and the lower Hudson River region carrying aquatic organisms in moist, solid ballast (and after the 1880s, water ballast) in their holds, human and livestock food stuffs, and fouling and boring organisms on and in their hulls. This vector contributed greatly to the introduction of non-indigenous species in the Hudson River. The opening of the Champlain Canal in 1819, the Erie Canal in 1825, and the creation, improvement, and expansion of many other smaller regional canals led to greatly increased shipping activity in New York by the mid-nineteenth century, as the Hudson River became one of the most important gateways into America's heartland (Adams 1981; Bruce 1907). The improvement of the New York State Erie Canal in 1918 and the modern renovation of the Port of Albany in 1932 accommodated larger sea-going vessels of 27 ft draft (Adams 1981). Construction of the Erie Canal further enhanced New York's strategic position as a major port system (Figure 2) and played a major role in facilitating the movement of fish species (Smith 1985).

Coincident with the development of agriculture and cities along the Hudson corridor were focused activities that led to the intentional introduction of non-native species. Plants were intentionally brought in for stock forage or for urban gardens and ponds. Fish were imported and released to improve local fisheries. With increased affluence and urban influence, the use of animals as aquarium pets increased. Intentional introductions often led to the secondary escape of these organisms or to their intentional release into the wild.

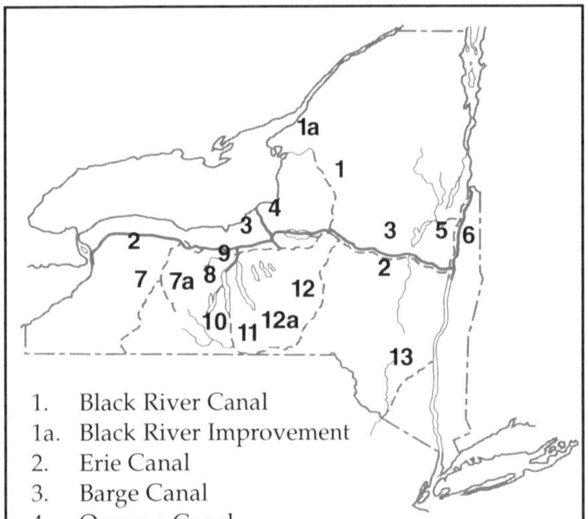

1. Black River Canal
1a. Black River Improvement
2. Erie Canal
3. Barge Canal
4. Oswego Canal
5. Glens Falls Feeder
6. Champlain Canal
7. Genesee Valley Canal
7a. Dansville Branch
8. Crooked Lake Canal
9. Cayuga and Seneca Canal
10. Chemung Canal
11. Junction Canal
12. Chenango Canal
12a. Chenango Extension Canal
13. Delaware and Hudson Canal

Figure 2. Canals of NewYork State.

Methods

We have followed the methods of Mills et al. (1993) and defined exotic species as successfully reproducing organisms transported by human-related activities into areas in which they did not previously exist. Also as in Mills et al. (1993), we have been forced to restrict our analyses of the non-indigenous biota to aquatic plants (macrophytes), fish, and, among the invertebrates, largely to mollusks and crayfish. The absence of historical information, the limitations of biogeographic studies, and a severe dearth of modern faunistic and systematic studies for many taxa in the Hudson River basin curtailed the analysis of the non-molluscan, non-crayfish exotic invertebrate biota of the basin. Data collection and analysis addressed the first date and location of introduction or collection, the mechanism of introduction, and the geographic origin of each taxon considered. Introduction mechanism definitions and codes are provided in Table 1.

Dates of Introduction

Dates of introduction are reported as

(1) the date of first recorded release (usually based upon a literature record),
(2) the date of the first recorded collection (based upon literature or museum records, or upon a personal communication), or
(3) the date of first publication.

Table 1. Introduction mechanisms for non-indigenous aquatic fauna and flora in the Hudson River basin (from Mills et al. 1993).

Mechanism	Code	Description
Deliberate	R(D)	The intentional release of an organism with the intention of establishing a population
Unintentional		The release of organisms without intention of creating established populations
Aquarium	R(AQ)	The unintentional release of aquarium pets or plants
Cultivation	R(C)	The accidental escape of cultivated plants from gardens or agriculture
Fish	R(F)	The unintentional release of organisms with unused bait or with other fish being stocked
Accidental	R(A)	The accidental introduction of organisms in any other manner
Shipping		
Fouling	S(F)	The transport of organisms attached to hulls of ships
Solid Ballast	S(SB)	The transport of organisms with the solid ballast of ships
Ballast Water	S(BW)	The transport of organisms within the ballast water of ships
Canals	C	The dispersal of organisms between areas through canals

We use first "sighting" and first "discovery" as synonyms of first collection. In the literature, first "recording" has been used to refer to both Categories 2 and 3, which we distinguish here. Examples of Category 3 are the records of DeKay (1842) and Goode (1903). In these and similar cases, we assume that the actual date of release or collection is prior to the date of publication, since it is unlikely that the first actual record is in fact the same date as the monograph in question. Thus, we always indicate these dates as "prior to" ("<") the date of publication. In those instances in which the literature does not provide precise dates or locations, referring instead to a general period (e.g., "mid-1800s") or a general location (e.g., "eastern New York"), we have attempted to ascertain a more specific date or location. In most cases, the first collection (Category 2) is sometime after the actual date of introduction into the river basin.

Mechanisms of Introduction

The mechanism or probable mechanism of introduction (entry vector) is defined as the known or the most probable means by which an organism gained access to the river basin. We have assigned an entry vector to each species based upon literature records, communications with relevant workers, and by analysis of the species' habitat. In addition, we have used known or suspected dispersal vectors from related taxa to infer a probable vector for those species for which a mechanism of introduction is unknown. Thus, if several species in a genus were introduced by escaping from cultivation, we suggest that an additional congener may have been introduced by the same means. For some species, the mechanism of introduction remains unknown or multiple mechanisms may apply, as noted below.

We divide introduction mechanisms into the following five broad groupings (Table 1):

- deliberate
- unintentional
- shipping activities
- canals, and
- multiple *(more than one possible introduction mechanism)*

We have assigned letter codes to each of these for use in the summary tables. "Deliberate introduction" refers to the intentional stocking of organisms into an area in hopes of establishing a viable population. An example is the stocking of black bass (*Micropterus* sp.) into interior waters of New York during the late 1800s (Cheney 1897a). An "unintentional introduction" is defined as the release of an organism without the intention of creating established populations. These include the introduction of aquarium animals and plants, the accidental escape of plants from cultivation, the release of unused bait fish with stocked fish, and the accidental release of organisms in any other manner (Carlton 1992a, 1993). "Shipping activities" include fouling organisms on ships' hulls, the transport of solid ballast, and the movement of water ballast (Carlton 1989). "Canals" refers to the natural movement of organisms through canals (as opposed to on barge hulls, which are included among shipping activities). Finally, for a number of species, multiple mechanisms have played a role. Further descriptions of these mechanisms have been provided by Mills et al. (1993).

Geographical Origin

Regional, continental, and subcontinental categories are used to identify the geographic origin of each taxon. These are

- Holarctic
- Eurasia
- Europe
- Asia
- North Africa
- North American Atlantic Coast
- North American Pacific Coast
- Southern United States
- Upper Mississippi
- Interior Basin
- Old World, and
- Unknown

Although "North American Atlantic Coast" covers the region from Florida to northern Canada, as used here it means a source region outside the Hudson River basin. Thus, species that are believed to have naturally occurred only to the north or to the south of the Hudson River system, but whose modern-day occurrence in the Hudson is due to human-mediated transport, are scored as originating in the North Atlantic region. "Southern United States" refers to waters either in the interior south or to the Gulf of Mexico. "Interior Basin" is a broad category that includes the area between the Rocky Mountains and the Appalachian Mountains. Some exotic species were introduced from a secondary origin. For example, the snail *Cipangopaludina chinensis malleata* is native to Asia, but it was almost certainly transported to the Hudson River basin from the Pacific Coast, where it was first introduced into North America (Mills et al. 1993).

Results

Aquatic Fauna

CHORDATA

Studies of the fishes of New York have a long historical tradition with a large body of literature associated with this work. Smith (1985) provided an excellent historical synopsis of Hudson River fishes. The earliest monographs to include the fishes of the Hudson River basin were Samuel Mitchill's reports (Mitchill 1815, 1817) on fishes of New York from 1814 to 1817 and James E. DeKay's *Zoology of New York* or *The New York Fauna; Part IV. Fishes*, published in 1842. Tarleton H. Bean's (1903) *Catalogue of the Fishes of New York* was the first study to separate the Hudson River from other locations in the state. The first comprehensive surveys of the Hudson River basin fish fauna were undertaken by the New York State Conservation Department in 1932-1936 (Greeley 1935, 1937; Greeley and Bishop 1933).

After almost 40 years, the need for a more recent survey was realized, and Bath et al. (1976) conducted a survey of the main channel of the Hudson River during the summer of 1976. Several recent works have further documented the Hudson River fish fauna (Beebe and Savidge 1988; Smith 1985; Smith and Lake 1990). The names of fish listed below follow those given by Robins et al. (1991). A summary of introduced fish in the Hudson River basin is presented in Table 2. We have also included a list of fish species with rare sightings in the Hudson (Table 3), which we consider as potential exotics. However, we have not included these in Table 2 because we do not consider them established.

Table 2. Origins, date and location of first sighting, and entry mechanism(s) for non-indigenous aquatic fauna of the Hudson River basin.

Taxon	Species	Common Name	Origin	Date	Location	Mechanism
Fish						
Petromyzontidae	*Ichthyomyzon unicuspis*	silver lamprey	Interior Basin	1974	LH, UH	C
Amiidae	*Amia calva*	bowfin	Interior Basin	<1988	LH	C
Clupeidae	*Dorosoma cepedianum*	gizzard shad	Interior Basin	1972	MH, LH	C
Ictaluridae	*Noturus miurus*	brindled madtom	Interior Basin	<1985	MH	C
Cyprinidae	*Campostoma anomalum*	central stoneroller	Interior Basin	1976	MH	C
	Carassius auratus	goldfish	Asia	<1832	MH, LH	R(D),R(AQ),R(A)
	Ctenopharyngodon idella	grass carp	Asia	1992	LH	R(A)
	Cyprinus carpio	common carp	Eurasia	1832	MH, LH, UP	R(D)
	Nocomis biguttatus	hornyhead chub	Interior Basin	<1900	MH	C
	Notropis atherinoides	emerald shiner	Interior Basin	<1934	MH, LH	C
	Notropis ludibundus	sand shiner	Interior Basin	1977	LH	R(F)
	Pimephales promelas	fathead minnow	Atlantic Coast	unknown	MH, LH	R(A)
	Rhodeus sericeus	bitterling	Eurasia	<1923	LH	R(D)
	Scardinius erythrophthalmus	rudd	Eurasia	1916	MH, LH	R(D)
Salmonidae	*Oncorhynchus mykiss*	rainbow trout	Pacific Coast	1874	UH, MH, LH	R(D)
	Salmo trutta	brown trout	Eurasia	<1887	UH, MH, LH	R(D)
Umbridae	*Umbra limi*	central mudminnow	Interior Basin	<1934	MH, LH	R(D),R(A)
Esocidae	*Esox lucius*	northern pike	Northern N. Am.	<1850	UH, MH, LH	R(D)
	Esox lucius x masquinongy	tiger muskellunge	Northern N. Am.	1980	MH	R(D)
Poeciliidae	*Gambusia affinis holbrooki*	eastern mosquitofish	Atlantic Coast	1992	LH	R(A)
Moronidae	*Morone chrysops*	white bass	Interior Basin	1975	MH, LH	C
Centrarchidae	*Ambloplites rupestris*	rock bass	Interior Basin	>1827	UH, MH, LH	C
	Lepomis cyanellus	green sunfish	Interior Basin	1936	MH, LH	R(D)
	Lepomis gulosus	warmouth	Interior Basin	1936	LH	R(A)
	Micropterus dolomieui	smallmouth bass	Interior Basin	>1827	UH, MH, LH	C, R(D)
	Micropterus salmoides	largemouth bass	Interior Basin	>1827	UH, MH, LH	C, R(D)
	Pomoxis annularis	white crappie	Interior Basin	<1934	MH, LH	C, R(D)
	Pomoxis nigromaculatus	black crappie	Interior Basin	>1903	MH, LH	C, R(D)
Percidae	*Percina caprodes*	logperch	Interior Basin	<1932	MH, LH	C
	Percina peltata	shield darter	Atlantic Coast	1936	LH	C
	Stizostedion vitreum vitreum	walleye	Interior Basin	1893	MH, LH	R(D)
Sciaenidae	*Aplodinutus grunniens*	freshwater drum	Atlantic Coast	1989	MH, LH	C
Mollusca						
Valvatidae	*Valvata piscinalis*	European stream valvata	Eurasia	unknown	LH	C, S(SB)

Table 2 continued.

Taxon	Species	Common Name	Origin	Date	Location	Mechanism
Bithyniidae	*Bithynia tentaculata*	mud bithynia	Eurasia	1879	MH	R(D)
Viviparidae	*Cipangopaludina chinensis*	Chinese mystery snail	Asia	1920	LH	R(AQ), R(A)
	Viviparus georgianus	banded mystery snail	Interior Basin	1867	MH	R(D)
Pleuroceridae	*Elimia livescens*	liver elimia	Interior Basin	1872	LH	C
	Pleurocera acuta	sharp hornsnail	Interior Basin	1856	MH	C
Lymnaeidae	*Radix auricularia*	big ear radix	Eurasia	<1869	MH, LH	R(AQ), R(A)
Unionidae	*Alasmidonta marginata*	elktoe	Interior Basin	<1869	MH	C
	Anodonta embecilis	paper pondshell	Interior Basin	1856	MH	C
	Fusconaia flava	wabash pigtoe	Interior Basin	1877	MH	C
	Lampsilis ovata	plain pocketbook	Interior Basin	1983	LH	C
	Leptodea fragilis	fragile papershell	Interior Basin	1853	MH	C
	Proptera alata	pink heelsplitter	Interior Basin	unknown	UH	C
	Pyganadon grandis	giant floater	Interior Basin	unknown	MH, LH	C
Sphaeriidae	*Pisidium amnicum*	greater European pea clam	Eurasia	1983	LH	R(AQ, F, A)
	Sphaerium corneum	European fingernail clam	Eurasia	1972	MH, LH	unknown
Dreissenidae	*Dreissena polymorpha*	zebra mussel	Eurasia	1991	LH	S(F)
	Mytilopsis leucophaeata	dark false mussel	Atlantic Coast	1937	LH	S(BW)
Mactridae	*Rangia cuneata*	Atlantic rangia	Southern U.S.	1991	LH	R(A), S(BW)
Arthropoda						
Cambaridae	*Orconectes immunis*		Interior Basin	1906	MH, LH	R(A),C
	Orconectes obscurus		Interior Basin	1934	MH	C,R(A)
	Orconectes rusticus	rusty crayfish	Interior Basin	1968	MH, LH	R(A),R(D)
	Orconectes virilis		Interior Basin	1957	MH, LH, UP	R(A)
	Procambarus acutus acutus		Interior Basin	1936	MH, LH	R(D),R(A)
Amphipoda	*Gammarus daiberi*		Southern U.S.	<1969	MH, LH	C
Coelenterata	*Cordylophora caspia*		Eurasia	unknown	LH	S(F)
	Craspedacusta sowerbyi	freshwater jellyfish	Asia	1943	MH	R(AQ)
Annelida						
Oligochaeta	*Ripistes parasita*		Eurasia	1978	MH, LH	S(BW), R(AQ)

Location codes: LH = Lower Hudson, MH = Mohawk-Hudson, and UH = Upper Hudson watersheds as defined by the dashed lines in Figure 2. Introduction mechanism codes are given in Table 1.

Table 3. Rare occurrences of non-indigenous fishes in the Hudson River basin.

Species	Origin	Date	Location	Mechanism	Reference
Lepisosteus osseus Longnose gar	Central & Eastern N. Am.	1989	RM 70	C	Smith and Lake 1990
Ictalurus punctatus Channel catfish	Central & Eastern N. Am.	1976	RM 56	C	Smith and Lake 1990
Oncorhynchus nerka Kokanee salmon	Pacific N. Am.	1974	RM 66	C, R(C)	Reider 1979
Oncorhynchus tshawytscha Chinook salmon	Pacific N. Am.	1938	RM 42	R(D)	NYSM 738
Cyprinodon variegatus Sheepshead minnow	Atlantic N. Am.	1974	RM 66	R(AQ)	Smith and Lake 1990

(RM = river mile; The Battery at New York City is RM = 0). For introduction mechanism codes, see Table 1.

Cephalaspidomorphi–Petromyzontidae

Ichthyomyzon unicuspis . . . . .Silver lamprey

Silver lamprey occurs in the Great Lakes and in the St. Lawrence River, Mississippi, and Ohio River systems (Lee et al. 1980; Smith 1985). The first *Ichthyomyzon unicuspis* was recorded for the upper Hudson River near Coveville in 1937 and 1938 (NYSM 11618 and 11620). Later, silver lamprey was observed in the lower Hudson River on a traveling screen of the Niagara Mohawk Power Corporation at river mile 142 (228 km) near Albany, New York, in June 1974 (Reider 1979). George (1985) and Reider (1979) attributed its presence in the Hudson River to its passing through the Champlain Canal since Greeley (1930) found it moderately common in Lake Champlain where it is a native species.

Osteichthyes–Amiidae

Amia calva . . . . .Bowfin

Bowfin occurs throughout the central plains from Wisconsin to Texas, down the Mississippi and Ohio River basins, and up the Atlantic coast from Florida to southern Pennsylvania (Lee et al. 1980; Smith 1985). George (1981) indicated that bowfin had not been reported in the Hudson River as of 1980, but in 1988, it was present in the lower Hudson River (Smith and Lake 1990). It may have moved through the Champlain Canal where it has been established for at least 60 years (Greeley 1930), or it may have been introduced by people (Smith 1985).

Osteichthyes–Clupeidae

Dorosoma cepedianum . . . . .Gizzard shad

Currently, gizzard shad occurs in the Great Lakes, the Mississippi River drainage basin, the Hudson-Mohawk River systems, and Atlantic coastal drainages as far north as New York harbor and Long Island (DeKay 1842; Greeley 1935, 1937; Lee et al. 1980; Smith 1985). Dew (1973) first reported gizzard shad in the lower Hudson in October 1972, and George (1983) noted specimens were taken from Collins Lake, a tributary of the Mohawk River in Schenectady County in June 1976. George (1983) noted the incidence of larger, more mature fish in the Mohawk system and the absence of similar fish in the Hudson

River as evidence for their movement down through the Erie Canal. Smith and Lake (1990) believed its occurrence in the Hudson River resulted from its movement through the Erie Canal because it is common there and uncommon in the New York harbor area.

Osteichthyes–Ictaluridae
Noturus miurus . . . . .Brindled madtom
Brindled madtom ranges from Lake Erie and Lake Ontario down through the Ohio River basin and the lower one-half of the Mississippi River basin to the Gulf of Mexico (Lee et al. 1980; Smith 1985). Taylor (1969) did not list any records of brindled madtoms east of Oneida Lake. However, Smith (1985) listed several specimens taken near Lock 7 (Schenectady), and Smith and Lake (1990) reported specimens taken as far up the stream as Lock 9 (Rotterdam Junction) of the Erie Canal. Based on the eastward progression of collections over time (Smith 1985), we believe the brindled madtom moved into the Mohawk River and Hudson River basin via the Erie Canal.

Osteichthyes–Cyprinidae
Campostoma anomalum . . . . . .Central stoneroller
Central stoneroller is widely distributed in the United States from Wisconsin to New York and south into Georgia and Mexico (Lee et al. 1980; Smith 1985). On 7 October 1976, an American Museum of Natural History survey captured a central stoneroller in Schoharie Creek (Smith and Lake 1990), a tributary of the Mohawk River. Smith and Lake (1990) reported it as quite common there now, and they attributed its presence in the Hudson River basin to canals with access to the Great Lakes.

Osteichthyes–Cyprinidae
Carassius auratus .Goldfish
Goldfish is native to eastern Asia (Lee et al. 1980) and was widely introduced into North America prior to 1832 (Smith 1985). By 1842, it was known to be in the Hudson River (DeKay 1842). It was also introduced into the United States from Japan in 1874 by Admiral Daniel Ammon of the United States Navy (Hervey and Hems 1948). Initially imported as ornamental fish, many goldfish had been released into private ponds by the turn of the century (Lachner et al. 1970), and it was the first exotic fish species to establish a reproducing population in open waters of North America (Courtenay and Stauffer 1990). Goldfish have been introduced into the Hudson River by a wide variety of transport vectors including intentional introductions by the U.S. Fish Commission, hatchery and outdoor pond escapees, and aquarium and unused bait releases (Lee et al. 1980).

Osteichthyes–Cyprinidae
Ctenopharyngodon idella . . . . .Grass carp
Grass carp is endemic to eastern Asia from the Amur River Basin to the West River and southward to the Chu River (Greenfield 1974; Lee et al. 1980). Grass carp was originally considered for importation as a possible means for controlling aquatic macrophytes (Swingle 1957). It was first introduced into the United States in 1963 at Auburn University in Alabama (Lee et al. 1980) and soon spread rapidly by researchers, by agency stockings aimed at reducing aquatic macrophytes, through interstate importation by private hatcheries, and via escapes from hatching sites. Grass carp has since been banned from non-regulated importation and release in 35 states, including New York, because of its potential harmful impacts on aquatic environments (Smith 1985). Two ponds in the Hudson River basin have been stocked with grass carp from private hatcheries (Smith 1985). In June 1992, three grass carp were caught in the Hudson River near river mile 113 (182 km), and shortly thereafter, another grass carp was caught in the same area (Walt Keller, New York State

Department of Environmental Conservation, personal communication).

Osteichthyes–Cyprinidae
Cyprinus carpio ....Common carp
Common carp is native to Europe and temperate Asia (Smith 1985). DeKay (1842) reported that the common carp was first introduced into the United States by a private citizen, Henry Robinson, in 1832 when he brought them here from France. After cultivating them in his ponds "for several years," he began intentionally releasing from one to two dozen carp every spring into the Hudson River near his residence in Newburgh. At some point after that, Mr. Robinson stated, "They have increased so much that our fishermen frequently take them in their nets" (DeKay 1842). Common carp has since been introduced into many areas of North America (Lee et al. 1980; Smith 1985) and is considered detrimental because it destroys aquatic vegetation and stirs up the bottom so the water is continually turbid (Smith 1985).

Osteichthyes–Cyprinidae
Nocomis biguttatus ....Hornyhead chub
Hornyhead chub ranges from western New York through the Great Lakes and down the Mississippi and Ohio River basins (Lee et al. 1980; Smith 1985). Bean (1903) was the first to list it for the Hudson River basin, but gave no specific collection dates. George (1981) and Smith (1985) indicated that hornyhead chub probably came through the Erie Canal.

Osteichthyes–Cyprinidae
Notropis atherinoides ....Emerald shiner
Emerald shiner occurs in southern Canada from the St. Lawrence River to the Northwest Territory and south through the Mississippi River basin to the Gulf of Mexico (Lee et al. 1980; Smith 1985). Its occurrence in the Hudson River may be due to access through the Erie Canal (George 1981; Smith 1985; Smith and Lake 1990; Snelson 1968). Greeley (1935) listed the emerald shiner in the Mohawk River and Hudson River (NYSM 28341 and 28438) in the 1934 survey, but it was not found below Catskill in the Hudson River during the 1936 survey (Greeley 1937). Beebe and Savidge (1988) noted that the emerald shiner was common in the lower Hudson River in later surveys, and if it had colonized the Hudson-Mohawk system after the last glacial period, it should have been collected during the 1936 survey.

Osteichthyes–Cyprinidae
Notropis ludibundus ....Sand shiner
The sand shiner is widespread through the central United States and southern Canada from the St. Lawrence River to the Rocky Mountains and south to Texas (Lee et al. 1980; Smith 1985). Smith (1985) reported an isolated introduced population (before 1985) in the Shawangunk Kill in southern New York. Its absence in the Delaware drainage (Smith and Lake 1990) indicates that it most likely did not move through the Delaware-Hudson Canal into the Hudson River. This exotic fish might well be a bait-bucket introduction (George 1981). This species is referred to as *N. stramineus* in the literature. Mayden and Gilbert (1989) demonstrated that it correctly is named *N. ludibundus*.

Osteichthyes–Cyprinidae
Pimephales promelas ....Fathead minnow
Smith (1985) indicated that the fathead minnow occurs throughout most of the middle of North America and has been introduced on the Atlantic Slope and west of the Continental Divide. The fathead minnow is commonly used as fish bait and probably was introduced to the Hudson drainage by the bait bucket.

Osteichthyes–Cyprinidae

Rhodeus sericeus .Bitterling

Bitterling is native to northern Europe and central and northeastern Asia (Lee et al. 1980; Smith 1985). On 16 September 1923, Dr. W. C. Kendall captured some bitterling in the Saw Mill River, Tarrytown, Westchester County (Dence 1925; Myers 1925). Bade (1926) reported a bitterling taken from a small brook near Tarrytown approximately two years earlier. He stated, "It is very probable that, about ten years ago when the hobby of raising and breeding fish had its humble beginning, a fancier imported a number of these fish, placing them into this tiny stream where they have been able to multiply unhindered." Schmidt et al. (1981) argued that since the only breeding populations of the bitterling in North America occur in two New York rivers, they must be the result of a single introduction prior to 1923. Courtenay and Stauffer (1990) have also classified the bitterling as an aquarium release.

Osteichthyes–Cyprinidae

Scardinius erythrophthalmus . . . . .Rudd

Rudd is native to Europe and western Asia (Lee et al. 1980; Smith 1985) and was first found in the United States in 1897 in Central Park, New York City (Bean 1897, 1903). It was introduced into the Roeliff-Jansen Kill in southeastern New York and was established there by 1936 (Greeley 1937). The American Museum of Natural History had specimens sent to them by Mr. Clive Brown on 11 November 1916 when he was at Unity Lodge, Copake Lake, Columbia County. It is likely that Mr. Brown released the rudd in the Copake Lake region, possibly both in the lake and in the stream (C. L. Smith, AMNH, personal communication). Courtenay and Stauffer (1990) listed rudd as an aquarium trade release. The release of rudd has also likely occurred with discarded fish bait because many New York bait dealers in the Hudson River basin had rudd mixed with the golden shiners they were selling as bait during the spring and summer of 1990 (Paul Geoghegan, Normandeau Associates Incorporated, personal communication). Because rudd can hybridize with native golden shiners (*Notemigonus crysoleucas*), the genetic pool of golden shiners will be altered (Anonymous 1993; Burkhead and Williams 1991). The rudd is currently very common in the Hudson River drainage (Keller, New York State Department of Environmental Conservation, personal communication).

Osteichthyes–Salmonidae

Oncorhynchus mykiss . . . . .Rainbow trout

Rainbow trout is native to the Pacific Ocean drainages of North America and Asia from northern Mexico to Alaska (Lee et al. 1980; Smith 1985). It was introduced into New York in 1874 when Seth Green transferred eggs from the McCloud River in northern California to his hatchery in Caledonia, New York (MacCrimmon 1971; Wales 1939). These eggs were subsequently hatched and the fry distributed to many areas within the Hudson River basin. As a result of this and many subsequent introductions, rainbow trout can now be found throughout the Hudson River basin wherever there is suitable habitat (Smith 1985).

Osteichthyes–Salmonidae

Salmo trutta .Brown trout

The brown trout is an Old World species ranging throughout Europe, western Asia, and northern Africa (Lee et al. 1980; Smith 1985). The first brown trout to enter the United States came from eggs sent from Germany in 1883 and were delivered to the Northville Hatchery, Michigan (Goode 1903; Mather 1889a). Subsequent shipments of eggs made between 1884 and 1887 were raised at the United States Fish Commission station in Caledonia, New York, and these fish were stocked in various New York waters including some streams in the Hudson River basin (Goode 1903). Goode (1903) did not specify a date of first release; we arbitrarily choose 1887. Brown trout is now abundant in Hudson River tributaries (Smith 1985; Smith and Lake 1990).

Osteichthyes–Umbridae

Umbra limi . . . . .Central mudminnow

The central mudminnow occurs from southern Canada, the Great Lakes and the Upper Mississippi River to the St. Lawrence River and south into Tennessee and Arkansas (Lee et al. 1980; Smith 1985). In New York it is abundant in the Erie Canal (Smith and Lake 1990), and since 1967, it has been collected at several locations within the Hudson River (Beebe and Savidge 1988). The central mudminnow most likely moved into the Mohawk River from the Great Lakes via the Erie Canal (Smith 1976, 1985; Smith and Lake 1990). It was first found in the Hudson River basin in the 1934 Mohawk River survey; Greeley (1935) indicated that it probably came through the Erie Canal. The central mudminnow is now common in the Hudson River basin, and since it is often used as bait (George 1981), it may have been introduced into other areas of the river through bait-bucket releases.

Osteichthyes–Esocidae

Esox lucius .Northern pike

Northern pike is a holarctic species generally restricted to waters north of a line from the St. Lawrence River through the Great Lakes and upper Mississippi and Missouri Rivers (Lee et al. 1980; Smith 1985). Northern pike is native to Lake Champlain and Lake George in northern New York but is considered non-native to the Hudson River (George 1981). During the 1840s, Sandford Lake and Henderson Lake, located in Essex County on the upper Hudson River, were stocked with northern pike taken from Schroon Lake, which had been previously stocked from Lake Champlain (Clarke 1852; George 1981). Northern pike was found in the upper Hudson River basin in 1932 (Greeley 1935) and in the Mohawk River in 1934 where it was believed to be an Erie Canal immigrant (Greeley 1935). It was collected in the lower Hudson River in 1976 (Bath et al. 1976). Smith (1985) and Smith and Lake (1990) listed several records of northern pike in tributaries of the Mohawk and upper Hudson Rivers.

Osteichthyes–Esocidae

Esox lucius x
Esox masquinongy . . . . .Tiger muskellunge

Tiger muskellunge, also known as norlunge, is a hybrid of northern pike and muskellunge. It has been typically stocked as game fish because of its sterility and rapid growth potential (Smith 1985). The New York State Department of Environmental Conservation has been stocking tiger muskellunge in the Mohawk River since 1980 (Norm McBride, New York State Department of Environmental Conservation, personal communication), from which it has moved downstream into the Hudson River. Smith and Lake (1990) listed one specimen caught as far south as river mile 29 (47 km) of the Hudson River in April 1987. Although tiger muskellunge is sterile and therefore cannot reproduce, its continued stocking in these waters merits attention because it can reduce native fish stocks through predation.

Osteichthyes–Poeciliidae

Gambusia holbrooki . . . . .Eastern mosquitofish

Eastern mosquitofish occurs in the Atlantic and Gulf Slope drainages and west to southern Alabama (Page and Burr 1991). In July 1992, the eastern mosquitofish was discovered in Sparkill Creek near the New Jersey border and in Sparkill's skating rink located near the creek's mouth in the Hudson River (Hamilton and Schmidt 1992); this was the first recorded capture in the Hudson River drainage. Robert Schmidt (Simons Rock College, personal communication) believes that it may have escaped from one or more ponds on a golf course that is upstream from their capture site. He also indicated that this fish could have been unintentionally released from nearby New Jersey where they are stocked for mosquito control. Either of these scenarios can be classified as an unintentional release.

Osteichthyes–Moronidae
Morone chrysops . . . . White bass
Lee et al. (1980) and Smith (1985) described the range of white bass as mid-continental, spanning from the St. Lawrence River through the Great Lakes to the Red River of the North and down the Mississippi and Ohio River basins. The 1934 New York State Biological Survey found white bass in Oneida Lake but nowhere east of there in the Hudson-Mohawk system (Greeley 1935, 1937). Beebe and Savidge (1988) noted that white bass has been collected in the Hudson River since 1975, and they believed it probably moved out of Oneida Lake through the Erie Canal. Texas Instruments caught a white bass at river mile 42 (68 km) of the Hudson River on 18 January 1976 (Smith and Lake 1990). Smith (1985) and Smith and Lake (1990) have also suggested that the presence of the white bass in the Hudson River has resulted from its movement through the Erie Canal. White bass is common in the Hudson River as over a dozen lots were catalogued in NYSM records between 1976 and 1979.

Osteichthyes–Centrarchidae
Ambloplites rupestris . . . . Rock bass
The original range of the rock bass has been greatly extended due to introductions (Trautman 1981). Lee et al. (1980) and Smith (1985) indicated that the rock bass occurs from New England through southern Canada and south through the Mississippi and Missouri River basins and east into the Tennessee River system. DeKay (1842) stated, "This species occurs abundantly in the Great Lakes, and the larger streams in the western counties of the state. Since the completion of the Erie and Champlain canals, it has made its appearance in the Hudson River." Rock bass was collected throughout the Hudson River basin during the 1932 (Greeley and Bishop 1933), 1934 (Greeley 1935), and 1936 (Greeley 1937) New York State Biological Surveys. Presently, it is abundant in freshwater habitats in all parts of the Hudson River basin (Smith 1985; Smith and Lake 1990).

Osteichthyes–Centrarchidae
Lepomis cyanellus . . . . Green sunfish
Green sunfish originally ranged through east-central North America including the Great Lakes basin west of the Appalachian Mountains and the Mississippi River drainage (Lee et al. 1980; Smith 1985). In New York, the green sunfish was initially found only in the Great Lakes drainage, but it has since been introduced into the New Croton Reservoir (Greeley 1937). During July and August of 1936, green sunfish was caught in Campfire Creek Pond near Millwood (NYSM records; Smith and Lake 1990) where it was stocked in hopes of establishing a sport fishery (Greeley 1937). The species apparently spread from there into the main channel of the Hudson River, although no records appear to exist prior to 1976 (Bath et al. 1976).

Osteichthyes–Centrarchidae
Lepomis gulosus . . . . Warmouth
Historically, warmouth ranged from Wisconsin and Ohio west to southern Pennsylvania and south down the Mississippi basin to Texas and east to Florida (Lee et al. 1980; Smith 1985). According to Greeley (1937), the first warmouth to enter the Hudson River basin was accidentally introduced with a shipment of green sunfish (*Lepomis cyanellus*) that was stocked in Campfire Creek Pond in 1936. It has also been introduced into Woodbury Creek, a tributary of Moodna

Creek, and in the Saw Kill near Tivoli (Smith and Lake 1990). Subsequently, warmouth was caught in August 1936 in the Saw Kill near Annandale (Smith and Lake 1990). It was collected again in the Hudson River in 1976 (Bath et al. 1976).

Osteichthyes–Centrarchidae
Micropterus dolomieu . . . . .Smallmouth bass
Micropterus salmoides . . . . .Largemouth bass
Early accounts of these fishes (e.g., Cheney 1897a; DeKay 1842; Seymour et al. 1869) often treated them as one species, the black bass. We address them together with respect to their invasions of the Hudson River basin. Smallmouth bass and largemouth bass originally ranged throughout the north-central part of the United States and southern Canada, the St. Lawrence drainage and the Mississippi basin, and also extended into Mexico and Florida (Lee et al. 1980; Smith 1985). In the beginning of the nineteenth century there were no black bass in any interior waters of New York (Cheney 1897a). Seymour et al. (1869) listed the Champlain Canal as the means through which black bass entered the Hudson River. However, according to Cheney (1897a), the completion of the Erie Canal in 1825 "brought black bass from Lake Erie to the Hudson River and to the waterways having connection with the canal." The black bass that entered the Erie Canal were subsequently used to stock other lakes in New York, many of which are in the Hudson River basin (George 1981). By 1903, both smallmouth and largemouth bass were collected in the Hudson River basin (Bean 1903). Greeley and Bishop (1933) reported that both species were present in the upper Hudson in 1932; they were also present in the Mohawk in 1934 (Greeley 1935). At present, both fish occur throughout the Hudson River basin; the smallmouth bass usually prefers cool, clear water with rocky substrate whereas the largemouth bass prefers warm, weedy areas (Smith 1985).

Osteichthyes–Centrarchidae
Pomoxis annularis . . . . .White crappie
White crappie occurs from southern Ontario and western New York to the Mississippi River basin and south to the Gulf of Mexico (Lee et al. 1980; Smith 1985). It was first collected in 1934 at sites in the lower Mohawk River (Greeley 1935). Greeley (1935) believed that white crappie was stocked in the Hudson-Mohawk system, but George (1981) believed, based on its original Mississippian range, that it entered through the Erie Canal. Smith and Lake (1990) also listed the white crappie as a canal immigrant now common in the Mohawk River. White crappie was captured in the Hudson River in 1976 (Bath et al. 1976).

Osteichthyes–Centrarchidae
Pomoxis nigromaculatus . . . . .Black crappie
Black crappie originally ranged from the St. Lawrence River basin west through the Great Lakes to the Mississippi and Missouri River basins, south to the Gulf of Mexico, and north up the Atlantic coast as far as northern Virginia (Lee et al. 1980; Smith 1985). Greeley (1935) indicated that black crappie was stocked in the Mohawk River basin in the early 1900s by the U.S. Bureau of Fisheries. This may have been after 1903, as Bean (1903) did not list it for the Mohawk River basin. However, Smith and Lake (1990) attributed its presence in the Hudson River basin to its passing through canals. The Erie Canal may have been the route, given original black crappie range in western and northern New York and its present distribution throughout the Hudson River (Smith 1985).

Osteichthyes–Percidae
Percina caprodes . . . . .Logperch
Logperch originally ranged from the Hudson Bay drainages south through the Great Lakes and the Mississippi River basin to the Gulf of Mexico (Lee et al. 1980; Smith 1985). Neither Bean (1903) nor DeKay (1842) mentioned the presence of the log-

perch in the Hudson River basin, although they both indicated that it was very common in Lake Champlain. Greeley (1935, 1937) and Greeley and Bishop (1933) listed it in the Hudson River basin with a first collection date of 1932. The logperch probably made its way into the Hudson River via the Champlain Canal.

Osteichthyes–Percidae
Percina peltata .Shield darter
The shield darter is native to Atlantic coastal drainages from the Delaware River south to North Carolina (Lee et al. 1980; Smith 1985). It was first collected in 1936 in Rondout Creek, a tributary of the Hudson River, by the New York State Biological Survey (Greeley 1937). The shield darter is common there now (Smith 1985). Rondout Creek was used as part of the passage for the Hudson-Delaware Canal, and *Percina peltata* likely gained access to Rondout Creek and the Hudson River basin through this canal (Beebe and Savidge 1988).

Osteichthyes–Percidae
Stizostedion vitreum vitreum . . . . .Walleye
Walleye ranges through the central part of North America from the MacKenzie River to Quebec and south through the Great Plains to northern Texas, Alabama, and Georgia, although its original distribution has been extended through introductions (Smith 1985; Trautman 1981). The Fisheries, Forest, and Game Commission of New York began stocking walleye into many lakes, rivers, and streams in 1893 (Cheney 1897b), several of which were either tributaries of the Hudson or Mohawk Rivers, or into the rivers themselves. The walleye has since become widespread throughout the state and is considered a valuable game fish.

Osteichthyes–Sciaenidae
Aplodinotus grunniens . . . . .Freshwater drum
Aplodinotus grunniens ranges throughout the Mississippi and Missouri basins, north to Montana and east to the Hudson Bay drainage and the St. Lawrence River (Smith 1985). Freshwater drum is common in big rivers and lakes, is generally restricted to shallow water, and is quite tolerant of turbidity (Smith 1985). In New York State, Smith (1985) indicated that freshwater drum is common in Lakes Ontario, Erie, and Champlain. Smith and Lake (1990) indicate a single sighting of freshwater drum in 1989. NYSM records include specimens from the Mohawk River (NYSM 41483) and the Hudson River at river mile 41 (NYSM 39600).

MOLLUSCA

The molluscan fauna of the Hudson River basin has benefited from a long history of research (e.g., Lewis 1856, 1860, 1872; Marshall 1895; Smith 1982, 1983; Townes 1937), but it was not until Strayer's (1987) paper that the fauna was treated as a whole. Our list builds upon that presented by Strayer (1987) and includes some recently invading mollusks. Exotic mollusks have entered the Hudson River basin through a number of vectors including canals, unintentional introductions, deliberate introductions, and via ships' ballast (Table 2). The common names of mollusks listed below follow those given by Turgeon (1988).

Gastropoda–Valvatidae
Valvata piscinalis . . . . . .European stream valvata
The European stream valvata was first collected in North America at Charlotte and Summerville, New York, near the mouth of the Genesee River on Lake Ontario in 1897 (Baker 1898). The first specimens known from the basin were collected by Strayer (1987) in 1985. *Valvata piscinalis* has been collected from various sites between Albany and Poughkeepsie along the tidal por-

tion of the Hudson River (Jokinen 1992; Strayer 1987). According to Latchford (1914, 1925, 1930), the snail was probably introduced into the Great Lakes within packaging used in ships transporting fragile merchandise from Europe. It could have also entered the Hudson River via the Erie or Champlain Canal.

Gastropoda–Bithyniidae
Bithynia tentaculata . . . . .Mud bithynia
Mud bithynia, also known as the faucet snail, a Eurasian native, was first found in North America by W. M. Beauchamp in 1879 in Oswego, New York, near Lake Ontario. Shortly thereafter, it was discovered in the Erie Canal near Syracuse (Beauchamp 1880). At some point not long after its discovery, James Lewis took some of the snails and transplanted them into the Mohawk River near Mohawk where they became very abundant (Gray 1883). In 1985, the species was very abundant on hard substrates in the lower Hudson River intertidal zone and now occurs in the Mohawk and Hudson Rivers (Strayer 1987). Jokinen (1992) also reported this species from Thompson Pond, Dutchess County, in the Hudson River basin, as well as from additional sites in the freshwater tidal Hudson.

Gastropoda–Viviparidae
Cipangopaludina chinensis . .Chinese mystery snail
The Chinese mystery snail was introduced into North America in San Francisco, California, in 1892 from Asia (Wood 1892) and is now widely distributed throughout North America (Clench and Fuller 1965; Jokinen 1982). It was discovered in the basin at in 1920 and is now known from various sites within the basin (Jokinen 1992; Strayer 1987). Strayer (1987) indicated that these snails may have been released into the Hudson River basin by amateur aquarists. Whether this snail found its way east by the 1920s as a food source is not known.

Gastropoda–Viviparidae
Viviparus georgianus . . . . . .Banded mystery snail
This snail is native to the Mississippi River basin and southeastern United States (Clench 1962; Clench and Fuller 1965) and is now widespread in lakes and rivers in the Hudson River basin (Simpson et al. 1984; Strayer 1987). James Lewis successfully introduced it in 1867 when he brought this snail from Illinois and planted it into the Erie Canal at Mohawk (Lewis 1872). This initial colony was very successful as the snails now have viable populations at various locations in the Mohawk and upper Hudson Rivers (Strayer 1987) and also lakes in the basin (Jokinen 1992). Its spotty occurrence in the basin suggests additional movements associated with aquarium introductions.

Gastropoda–Pleuroceridae
Elimia livescens
(=Goniobasis livescens) . . . . .Liver elimia
Early treatments (e.g., Goodrich 1942; Letson 1905) of the molluscan fauna incorrectly described this species as native to the Hudson River basin. *Elimia livescens*, native to both the Mississippi and Great Lakes basins, reached the Hudson River system from the Great Lakes through the Erie Canal; it was first collected in 1872 (Lewis 1872; Smith 1983). It now occurs mostly in larger stretches of the Hudson River basin, although it can occupy other habitats such as springs and lakes (Strayer 1987).

Gastropoda–Pleuroceridae
Pleurocera acuta . . . . .Sharp hornsnail
Pleurocera acuta is native to the Mississippi River and Great Lakes basins (Smith 1983). The earliest record of this species from the Hudson River basin is Lewis's (1856) mention of a specimen taken from the Mohawk River near Mohawk. This species likely entered the Hudson River basin through the Erie Canal (Goodrich 1942; Smith 1983). *Pleurocera acuta* has expanded its range eastward and now occurs in the main body of the Hudson River (Smith 1983; Strayer 1987).

Gastropoda–Lymnaeidae

Radix auricularia . . . . Big-ear radix

Big-ear radix, also known as the European ear snail, has been introduced into various areas of North America (Burch 1982; Clarke 1981) and first entered the Hudson River basin before 1869 (Strayer 1987). It is apparently uncommon in the Hudson River basin. Aldrich (1869) noted a few records of the big-ear near Troy, and Strayer (1987) found one specimen in a museum collection from Robinson Pond in Columbia County. Jokinen (1992) listed a record of the snail from a swamp near Cambridge, Washington County, and also indicated that it may have moved around with imported plants and lumber. Robertson and Blakeslee (1948) noted that this is a popular aquarium snail, and therefore it may have been an unintentional introduction associated with the aquarium trade.

Bivalvia–Unionidae

Alasmidonta marginata . . . . Elktoe

This Interior Basin species probably entered the Hudson River basin through the Erie Canal (Strayer 1987). The elktoe was common in the Mohawk River in the nineteenth century (Aldrich 1869; Strayer 1987) but has been rarely observed in the Hudson River basin in the twentieth century. In 1989, D. Strayer (unpublished data) found a fresh shell at the mouth of Schoharie Creek, a Mohawk River tributary near Fort Hunter.

Bivalvia–Unionidae

Anodonta imbecilis . . . . Paper pondshell

Strayer (1987) indicated a sighting of two specimens of this Interior Basin species in the Hudson basin in the late 1800s and suggested that it probably entered the basin via the Erie Canal. Its present status is unknown.

Bivalvia–Unionidae

Fusconaia flava . . . . Wabash pigtoe

Collections of the Wabash pigtoe, an Interior Basin species, were made in the Hudson River basin between 1877 and 1895. The present status is not known, but it almost certainly entered the Hudson basin through the Erie Canal (Strayer 1987).

Bivalvia–Unionidae

Lampsilis ovata . . . . Plain pocketbook

Strayer (1987) indicated that there is only one record of this unionid from the Hudson basin, and it probably entered the basin through the Champlain Canal. Since *Lampsilis ovata* is established in central New York, it could also have entered the Hudson basin from the Erie Canal.

Bivalvia–Unionidae

Leptodea fragilis . . . . Fragile papershell

According to Strayer (1987), *Leptodea fragilis* is an Interior Basin species that probably entered the Hudson basin via the Erie Canal. The status of this unionid in the Hudson basin is unknown.

Bivalvia–Unionidae

Proptera alata . . . . Pink heelsplitter

Proptera alata is an Interior Basin species that is established in Lake Champlain and central New York and rare to the Hudson basin (Strayer 1987). *Proptera alata* probably entered the Hudson basin through the Champlain Canal, the Erie Canal, or both.

Bivalvia–Unionidae

Pyganodon grandis . . . . Giant floater

Known from several sites along the mainstream Mohawk and Hudson Rivers, this interior basin clam most likely invaded the Hudson River basin through the Erie Canal, the Champlain Canal, or both (Strayer 1987). It is not known when *Pyganodon grandis* invaded the area. Strayer (1987) has argued that its restricted distribution reflects a relatively recent arrival. By the middle of the nineteenth century, Lewis (1860) found it "abundant in the Erie Canal." It can hybridize with its eastern counterpart, *Anodonta cataracta* (Kat 1986), and some of the populations of *Anodonta* in the basin may indeed be hybrids (Strayer 1987).

Bivalvia–Sphaeriidae

Pisidium amnicum . . .Greater European pea clam

This Eurasian clam was discovered in the Hudson River during the summer of 1983 when the New York State Department of Health found them in its macrobenthic survey at river mile 141 (227 km), 136.5 (220 km), and 134 (216 km) near Glenmont (Simpson et al. 1984, 1986). This clam was probably introduced into North America through the ballast water of trans-Atlantic ships since its early distribution in North America was restricted to the St. Lawrence River, the eastern Great Lakes, New Jersey, and eastern Pennsylvania (Burch 1975).

Bivalvia–Sphaeriidae

Sphaerium corneum . . . .European fingernail clam

The European fingernail clam has been found in the Great Lakes and in Lake Champlain (Herrington 1962). Simpson (1976) found the European fingernail clam in the Hudson River near Albany during his surveys in 1972 and 1973. However, Simpson et al. (1984) did not find any specimens of *S. corneum* during their extensive macrobenthic sampling in the summer of 1983. Strayer (1987) found a single empty shell in the Hudson River near Kingston and noted that the clam's presence in the Hudson River basin requires confirmation because these unionids are very difficult to identify. It is not known how *S. corneum* arrived in the Hudson River basin, although it could have gained access through the Erie Canal from the Great Lakes or with ballast water from trans-Atlantic ships.

Bivalvia–Dreissenidae

Dreissena polymorpha . . . . .Zebra mussel

This Eurasian mussel was first discovered living in North America in Lake Erie in December 1987 (Leach 1993) and again in Lake St. Clair in June 1988 (Hebert et al. 1989). *Dreissena polymorpha* was first found in the Hudson River in May 1991 near Catskill at river mile 112 (180 km) in small numbers (Powell and Strayer 1992; Strayer and Powell 1992). The zebra mussel was detected in the summer of 1992 as far south as West Haverstraw (river mile 37) (60 km) in salinities of 2 to 6 parts per thousand (Walton 1993). The zebra mussel has also moved east through the Erie Canal into the Mohawk River from the Great Lakes (Mills et al. 1993). It was discovered in Lake Champlain in the summer of 1993 (LaBar, University of Vermont, personal communication). Carlton (1993) has reviewed dispersal mechanisms within and between water bodies. Dispersal up river and up canals is largely vessel-mediated (as hull-fouling organisms or in ballast water), while dispersal down river is the result of a combination of natural planktonic dispersal and vessel traffic. Strayer and Powell (1992) suggested that the zebra mussel was first introduced near Catskill, since it did not occur at the time of discovery upstream of that site nor downstream of the Mohawk River mouth. Numerous environmental and societal impacts accompany the invasion of zebra mussels into new regions, including the fouling of water intakes and boats, the alteration of water column trophic pathways, and negative impacts on native biota, particularly native freshwater unionid mussels (Nalepa and Schloesser 1993).

As of September 1994, the second species of zebra mussel in North America (*Dreissena bugensis*) (Spidle et al. 1994) had not yet been detected in the Hudson River.

Bivalvia–Dreissenidae

Mytilopsis leucophaeata ....Dark false mussel

The dark false mussel occurs in brackish water (oligohaline) from the Gulf of Mexico to Chesapeake Bay (Marelli and Gray 1985). It may or may not be native to Chesapeake Bay (Carlton 1993). It has been found in three localities north of Chesapeake Bay: the Hudson River near Haverstraw (river mile 37) (60 km) for the first time in 1937 and again in 1952 (Jacobson 1953; Rehder 1937), and in the Housatonic River, Connecticut (D. Smith, University of Massachusetts, Amherst, personal communication). The dark false mussel is now common in the Hudson River (Crandall 1977; Ristich et al. 1977), at least as far north as Newburgh (river mile 60) (97 km) (Walton 1993, W. Walton, personal communication). This oligohaline mussel may have been transported into the Hudson River in the ballast water of coastal vessels, or it may have transferred into the river with commercial oysters.

Bivalvia–Mactridae

Rangia cuneata ....Atlantic rangia

The Atlantic rangia, or southern marsh clam, is an oligohaline species native to Gulf of Mexico estuaries (Pennak 1978). The range of *Rangia cuneata* was expanded, probably by human activity, to the Atlantic coast in the twentieth century, being recorded from Chesapeake Bay by 1955 (Hopkins and Andrews 1970). It remained unknown north of the Chesapeake until 1988, when it was discovered by Christopher Letts in the lower Hudson near Haverstraw (Carlton 1992c; C. Letts, personal communication). The new occurrence of Atlantic rangia in the Hudson River may be due to one or more of three potential dispersal mechanisms: (1) it may have been introduced through release of adults, which are often sold as food for human consumption (Charles deQuillfeldt, New York State Department of Environmental Conservation, personal communication); (2) it may have been introduced with the re-laying of commercial oysters (R. Everett, United States Fish and Wildlife Service, personal communication); or (3) it may have arrived in the ballast water of ships (Carlton 1992b).

Bivalvia–Corbiculidae

Corbicula fluminea ....Asian clam

As of September 1993, *Corbicula* was not known from the Hudson River basin. We note it here, however, since it occurs in both freshwater and brackish water in Chesapeake Bay to the south (Carlton 1993), and in freshwater in the Connecticut River, Connecticut, to the north (Morgan et al. 1991). Empty valves and living specimens of *Corbicula* are frequently encountered far from established populations, representing discarded fish bait. We thus expect that *Corbicula* will eventually be found in the Hudson River or one of its tributaries in the near future.

ARTHROPODA

Decapods are the best known of the freshwater crustaceans (Peckarsky et al. 1990). In northeastern United States, crayfish species occur in a wide variety of habitats, including streams, large rivers, ponds, lakes, swamps, and wet meadows (Peckarsky et al. 1990; Pennak 1978). They are generally omnivorous, feeding on succulent aquatic vegetation and invertebrates such as snails (Peckarsky et al. 1990; Pennak 1978). A list of non-indigenous crayfish is shown in Table 2.

Amphipods usually occur in the littoral benthos of unpolluted lakes, ponds, streams, streams, springs, and subterranean waters (Peckarsky et al. 1990; Pennak 1978). They are generally nocturnal omnivorous scavengers, feeding on microscopic plants, animals, and debris (Peckarsky et al. 1990; Pennak 1978).

Decapoda–Cambaridae

Orconectes immunis . . . . .Calico crayfish

There is debate as to whether *O. immunis* is introduced or native to the Hudson River basin. Crocker (1957, 1979) considered it native, whereas D. Smith (University of Massachusetts-Amherst, personal communication) believes it is introduced. This species is widely propagated for fish bait (Crocker 1979; Forney 1956), and therefore its distribution has almost certainly been expanded by human activities. A pre-European entry into the basin is possible, but the distribution presented by Crocker (1957) could easily be explained by human introductions (e.g., bait-bucket discards). The first record of *Orconectes immunis* in the Hudson River basin is given by Ortmann (1906) for Rensselaer Lake, Rensselaer County. *Orconectes immunis* is a slow-moving species, and it could have spread eastward through the Erie Canal from native Great Lakes populations.

Decapoda–Cambaridae

Orconectes obscurus

The first published record of this crayfish in the Hudson River basin is in 1934 for the upper Mohawk River (Nevin and Townes 1935). Crocker (1957) presents a disjunct distribution of *Orconectes obscurus* in the upper Mohawk River and states that the population may have arrived there via the Erie Canal. Other dispersal mechanisms could include bait-bucket introduction or an escape from propagation since there are very few records of the crayfish west of Oneida. Crocker (1957) also indicated that large numbers of *O. obscurus* were collected from the Erie Canal near Utica for use as fish bait in Adirondack lakes. This organism has also been introduced into the Hoosic River system, tributary to the Hudson River, in Massachusetts (Smith 1991).

Decapoda–Cambaridae

Orconectes rusticus . . . . .Rusty crayfish

A native of the Interior Basin, *Orconectes rusticus* was first found in New York in 1968 in an artificial pond about 6 miles west of Schenectady, Schenectady County (Crocker 1979), well within the Hudson River basin. Crocker (1979) attributed its presence there to "introductions by man." The rusty crayfish has also been introduced into areas as discarded bait (Capelli 1982). *Orconectes rusticus* is an aggressive species that will readily displace native species through direct competition for food and suitable habitat (Capelli 1982; Lodge et al. 1986; Olsen et al. 1991). It can also dramatically reduce aquatic macrophytes to the point of causing significant declines in centrarchid fish populations (Lodge and Lorman 1987; Lorman and Magnuson 1978). *Orconectes rusticus* is also established in the Titicus River, Westchester County, and Fairfield County in Connecticut (unpublished records, University of Massachusetts Museum of Zoology).

Decapoda–Cambaridae

Orconectes virilis . . . . .Virile crayfish

Orconectes virilis is native to the Mississippi River basin (Crocker 1957). Crocker (1957) noted its presence at two localities in the Hudson River basin: near Saratoga Springs and in eastern Westchester County. Crocker (1957) proposed a post-glacial spread of *Orconectes virilis*, although he did acknowledge the possibility of a food-industry escape based on Faxon's (1898) account of its transport from Milwaukee, Wisconsin, to New York markets. Later accounts also support its introduction by humans (Crocker 1979; Smith 1979), probably through the discarding of food waste or bait-bucket transfer. Smith (1979) gave several records in Columbia and Rensselaer Counties. Bovbjerg (1970) noted that *Orconectes virilis* will also displace other native species by outcompeting them for available food and habitat. *Orconectes virilis* is now everywhere in the Hudson drainage.

Decapoda–Cambaridae

Procambarus acutus acutus . .White river crawfish

The first record of this Mississippi River basin crayfish in New York was given by Crocker (1957) for the Bronx River based upon a 1936 New York State Department of Environmental Conservation survey and his own collections there in 1952. Pickett and Sloan (1985) provided a detailed synopsis of the introduction and spread of *Procambarus acutus acutus* in the northern part of the Hudson River basin. Beginning in the spring of 1951, several thousand crayfish were transported from southern Illinois to bait dealers in New York over a five-year period. These authors found it abundant in several New York lakes in Albany and Rensselaer Counties, and also reported it in Dutchess and Westchester Counties; all these counties are in the Hudson River basin. *Procambarus acutus acutus* will also displace and dominate native crayfish communities through competition for available food and suitable habitats (Pickett and Sloan 1985).

Amphipoda-Gammaridae

Gammarus daiberi

This freshwater amphipod is native along the Atlantic coast north to the Chesapeake and Delaware Bay region. Bousfield (1969, 1973) speculated that its occurrence north of this region, in particular in the Hudson River system, may be due to its spread through canals (Table 2).

COELENTERATA

Hydroids are sessile organisms and may be found attached to stones, twigs, vegetation, and debris in ponds, spring brooks, unpolluted streams, and the littoral zone of lakes (Pennak 1978). Many hydroids produce a planktonic dispersal stage, the hydromedusa. They possess minute stinging capsules, or cnidocytes, imbedded in epidermal cells that they use to paralyze their prey (Pennak 1978). Two non-indigenous coelenterates are listed in Table 2.

Cordylophora caspia (=C. lacustris)

This common Eurasian hydroid was transported to North America as a fouling organism on ships' hulls. Its presence by 1860 near Boston, Massachusetts (Verill 1874), and by 1870 in Philadelphia, Pennsylvania (Leidy 1870), suggests that this hydroid was established in lower Hudson River fresh and oligohaline waters by the same period, but there appear to be no early records. Ristich et al. (1977) recorded it in 1972 in the Hudson River, as did Crandall (1977) in 1974.

Craspedacusta sowerbyi . . . . . .Freshwater jellyfish

Native to Asia (Kramp 1950), the first North American record of *Craspedacusta sowerbyi* was in 1916 in Benson Creek, near Frankfurt, Kentucky (Garman 1916). *Craspedacusta* was first observed in North America in 1885 by Edward Potts who described the polyp (medusae was unknown) as *Microhydra ryderi*. Subsequent discovery of the medusa led to the realization that the two forms were different life stages of the same species. The first records for the Hudson River basin are for Garnet Lake, Warren County, and near Loudonville, Albany County, in 1943 (Deevey and Brooks 1943). Lytle (1960) also presented four records of *Craspedacusta sowerbyi* in the Hudson River basin, and Kiviat (Bard College, personal communication) has indicated that this organism has an established population in Silver Lake, Dutchess County. Based on its scattered distribution throughout the United States and its preference for artificial habitats, *Craspedacusta sowerbyi* may have been released with aquarium discharge or aquatic plants (Bushnell and Porter 1967).

ANNELIDA

Oligochaetes are common in most freshwater habitats, but are often ignored by freshwater biologists due to the perceived difficulty in identification (Strayer 1990). Recently, several publications (Brinkhurst 1986; Hiltunen and Klemm 1980; Klemm 1985; Simpson et al. 1984) have provided illustrated working keys to the North American freshwater tubificids, naidids, and lumbriculids making identification of oligochaetes relatively easier (Strayer 1990). Although we list only one exotic oligochaete in Table 2, there are certainly others which have yet to be identified.

Oligochaeta–Naididae
Ripistes parasita
This oligochaete, native to Europe (Brinkhurst and Jamieson 1971; Klemm 1985), has also been reported from New York (Brinkhurst 1986). The first records of this species in North America were obtained in 1978 in several New York rivers and in the New York canal system (Simpson and Abele 1984). Winnell and Jude (1987) suggested it was introduced into the Great Lakes in the ballast water of ships plying from Europe, although *Ripistes parasita* is an epiphytic species and therefore may also have been introduced with aquatic plants through the aquarium or horticultural trade.

Aquatic Flora

Northeastern North America was an active area of botanical research throughout the 1800s. New York in particular has had a rich history of floristic studies. John Torrey's (1843) *A Flora of the State of New York*, is the first and only completed flora of the State. Homer D. House, the second State Botanist of New York, concentrated primarily on flowering plants and produced two major publications: *Annotated List of the Ferns and Flowering Plants of New York State* (1924) and *Bibliography of the Botany of New York State* (1942). In 1947, Stanley J. Smith was appointed Curator of Botany of the New York State Museum and is credited with over two-thirds of the entries in the museum's plant distribution files. In 1962, he published *A Checklist of the Grasses of New York State* (Smith 1965). The present State Botanist, Richard S. Mitchell, has published *A Checklist of New York State Plants* (Mitchell 1986). In this study we define aquatic flora as those plants whose roots thrive in moist or wet habitats and whose shoots are submergent, floating, or emergent. Table 4 summarizes the data for plants.

In this study, we followed the plant taxonomic nomenclature of Gleason and Cronquist (1991), utilized published literature when possible, and did not seek information through the examination of museum records. Additionally, we focused on plants considered aquatic according to the criteria of Godfrey (1979, 1981). Many of the dates and locations for the following plants are listed on the basis of collections made by various amateur and professional botanists. In these instances, the date and location is followed by the collector's name, their collection number, if any, and an abbreviation for the herbarium name where the specimen has been placed.

Table 4. Origin, date and location of first sighting, and entry mechanisms for non-indigenous aquatic flora of the Hudson River basin.

Taxon	Habitat[1]	Species	Common Name	Plant Type[2]	Origin	Date	Location	Mechanism
Cabombaceae	A	*Cabomba caroliniana*	fanwort	p	Southern U.S.	1955	Yankee Pond	R(AQ)
Ranunculaceae		*Ranunculus acris*	common buttercup	p	Europe	1892	Highland	unknown
		Ranunculus bulbosus	bulbous buttercup	p	Europe	1896	Highland	unknown
		Ranunculus repens	creeping buttercup	p	Europe	1864	Schenectady Co.	unknown
	W	*Ranunculus scleratus*	cursed crowfoot	?	Eurasia	1863	Poughkeepsie	unknown
Polygonaceae	M	*Polygonum cespitosum* var. *longisetum*	bristly lady's-thumb	a	Asia	1936	Troy	unknown
	M	*Polygonum convolvulus*	black bindweed	a	Europe	1869	Poughkeepsie	? S(SB)
	W	*Polygonum hydropiper*	water-pepper	a	Europe	1926	Guilderland Center	unknown
	M	*Polygonum lapathifolium*	dock-leaved smartweed	a	Europe	1898	Highland	? S(SB)
		Polygonum persicaria	lady's thumb	a	Europe	1892	Highland	unknown
		Rumex crispus	curly dock	p	Europe	1924	Green Island	unknown
	M	*Rumex obtusifolius*	bitter dock	p	Europe	1927	Guilderland Center	unknown
Salicaceae		*Salix alba*	white willow	p	Europe	1851	Amenia	R(C)
		Salix fragilis	crack willow	p	Europe	1898	Yonkers	R(C)
		Salix purpurea	purple osier	p	Europe	1920	Albany	R(C)
Brassicaceae	W	*Cardamine pratensis*	cuckoo-flower	p	Eurasia	1864	Sander's Lake	unknown
	A	*Rorippa nasturtium-aquaticum*	water-cress	p	Eurasia	1893	Highland	R(C)
	W	*Rorippa palustris* var. *palustris*	common yellow-cress	a/b	Eurasia	1875	Pine Plains	S(SB)
	W	*Rorippa sylvestris*	creeping yellow-cress	p	Eurasia	1896	New Paltz	S(SB)
Primulaceae	M	*Lysimachia nummularia*	moneywort	p	Europe	1867	Saratoga	R(C)
		Lysimachia vulgaris	garden-loosestrife	p	Eurasia	1893	Ohioville	R(C)
Haloragaceae	A	*Myriophyllum spicatum*	European water-milfoil	p	Europe	1970	Constitution Island	R(AQ)
		Myriophyllum heterophyllum		?		1953	Kanawauke Lake	unknown
Lythraceae	W	*Lythrum salicaria*	purple loosestrife	p	Eurasia	1867	main channel	S(SB), R(C), R(A)
Trapaceae	A	*Trapa natans*	water chestnut	a	Eurasia	1884	Collins Lake	R(D)
Onagraceae	M/W	*Epilobium hirsutum*	hairy willow-herb	p	Eurasia	1866	Albany	S(SB)
Solanaceae	M	*Solanum dulcamara*	bittersweet	p	Eurasia	<1843	along main channel	R(C)
Menyanthaceae	A	*Nymphoides peltata*	yellow floating-heart	p	Europe	1929	Waterford	R(C)
Boraginaceae		*Myosotis discolor*	yellow and blue scorpoin grass	a	Europe	1882	Riverdale	unknown
	W	*Myosotis scorpioides*	water scorpion-grass	p	Europe	1868	Poughkeepsie	R(C)

Table 4 continued.

Taxon	Habitat[1]	Species	Common Name	Plant Type[2]	Origin	Date	Location	Mechanism
Lamiaceae		*Lycopus europaeus*	European water-horehound	p	Europe	1879	New York City	S(SB), R(C)
	M	*Mentha x gentilis*	red mint	p	Europe	unknown	along main channel	R(C)
	M	*Mentha x piperita*	peppermint	p	Europe	<1843	along main channel	R(C)
	M	*Mentha spicata*	spearmint	p	Europe	<1843	along main channel	R(C)
	M/W	*Stachys palustris* var. *palustris*	hedge-nettle	p	Europe	1867	Poughkeepsie	S(SB)
Scrophulariaceae		*Veronica serpyllifolia* var. *serphyllifolia*	thyme-leaved speedwell	p	Europe	1836	Troy	unknown
Rubiaceae	M	*Galium aparine*	cleavers	a	Eurasia	1927	Glenmont	unknown
Asteraceae	W	*Bidens tripartita*	beggar-ticks	a	Europe	1879	New York City	S(SB)
	M/W	*Boltonia asteroides* var. *recognita*	boltonia	p	Upper Mississippi	1929	Brewster	R(C)
	M	*Helenium flextuosum*	southern sneezeweed	p	S.E. U.S.	1904	Jonesville	unknown
		Sonchus arvensis	perennial-sow-thistle	p	Europe	1868	Poughkeepsie	S(SB)
Butomaceae	A	*Butomus umbellatus*	flowering rush	p	Eurasia	1965	Schagticoke	? C, R(D)
Potamogetonaceae	A	*Potamogeton crispus*	curly pondweed	p	Europe	<1814	Lake George	R(F), R(D)
Najadaceae	A	*Najas minor*	eutrophic water-nymph	?	Europe	1934	Waterford	unknown
Juncaceae	W	*Juncus inflexus*	rush	p	Eurasia & N. Africa	1917	Sangerfield	? R(C)
Poaceae		*Agrostis gigantea*	redtop	p	Europe	1924	Lower Jay	R(C)
	W	*Alopecurus geniculatus*	marsh-foxtail	p	Eurasia	1881	New York City	S(SB)
		Calamagrostis epigejos	feathertop	p	Eurasia	1930	Crescents	unknown
		Echinochloa crusgalli	barnyard-grass	a	Eurasia	1894	Highland	? R(C)
		Microstegium vimineum	Japanese stilt-grass	?	Asia	1987	Iona Island	unknown
	W	*Phragmites australis*	common reed	p	Eurasia	1844	Schenectady	? S(SB)
		Poa annua	speargrass	a	Eurasia	1921	Napanoch	? R(C)
		Poa nemoralis	wood bluegrass	p	Europe	1892	Davenport Place	? R(C)
		Poa trivialis	rough bluegrass	p	Europe	1811	New York City	? R(C), S(SB)
Iridaceae	W	*Iris pseudacorus*	yellow flag	p	Europe	1868	Poughkeepsie	R(C)

For introduction mechanism codes, see Table 1. Common name taken after Gleason and Cronquist (1991).

KEY

[1]A = aquatic, W = wet soil/swampy, M = moist soil

[2] a = annual, p = perennial, a/b = annual biennial

Our abbreviations are as follows:

- **BUF**, the Clinton Herbarium, Buffalo Museum of Science, Buffalo, New York
- **BH**, the Bailey Hortorium, Cornell University, Ithaca, New York
- **GH**, the Gray Herbarium, Harvard University, Cambridge, Massachusetts
- **HHH**, the Hoysradt Herbarium, Hartwick College, Oneonta, New York
- **MICH**, the University of Michigan Herbarium, Ann Arbor, Michigan
- **NY**, the New York Botanical Garden, New York, New York
- **NYS**, the New York State Museum Herbarium, Albany, New York
- **TRLSD**, Trailside Museum, Bear Mountain, New York
- **VAS**, the Vassar College Herbarium, Poughkeepsie, New York

Cabombaceae
Cabomba caroliniana . . . . .Fanwort
This common aquarium species was introduced into the northeastern and midwestern United States and Canada from the southeastern United States (Voss 1985). It was first recorded in the Hudson River basin in August 1955, in Yankee Pond near Woodstock, Ulster County (*H.F. Dunbar s.n.* NYS). *Cabomba caroliniana* was probably an accidental or unintentional release associated with the aquarium trade (Mills et al. 1993).

Ranunculaceae
Ranunculus acris . . . . .Common buttercup
Common buttercup is an European native now common in fields, pastures, and clearings (Fernald 1950; Gleason and Cronquist 1991). *Ranunculus acris* was first recorded in the Hudson River basin in August 1892, in Highland, Ulster County (*C.J. Elting 50* NYS). Its mechanism of introduction is unknown.

Ranunculaceae
Ranunculus bulbosus . . . . .Bulbous buttercup
This European plant (Fernald 1950) was first recorded in the Hudson River basin in May 1896, in Highland, Ulster County (*C.J. Elting 56* NYS). Bulbous buttercup grows in fields, pastures, and along roadsides and woods (Britton and Brown 1896; Fernald 1950; Gleason and Cronquist 1991; Gray 1848; Torrey 1843). Its mechanism of introduction is unknown.

Ranunculaceae
Ranunculus repens . . . . .Creeping buttercup
Paige (1864) gave the first record of *Ranunculus repens* in the Hudson River basin as 1864 from Schenectady County. Native to Europe, *Ranunculus repens* grows in wet, shady areas along rivers and ditches (Gray 1848; Torrey 1843) and is considered a rapidly spreading weed (Fernald 1950). Its mechanism of introduction is unknown.

Ranunculaceae
Ranunculus sceleratus . . . . .Cursed crowfoot
Introduced from Eurasia (Fernald 1950), *Ranunculus sceleratus* was first recorded in the Hudson River basin in July 1863, in Poughkeepsie, Dutchess County (*W.R. Gerard s.n.* NYS). Cursed crowfoot grows in low, wet areas such as ditches, sloughs, and swamps (Gleason and Cronquist 1991; Gray 1848; Torrey 1843) and prefers saline or alkaline conditions (Britton and Brown 1896; Fernald 1950). Its mechanism of introduction is unknown.

Polygonaceae
Polygonum cespitosum
var. *longisetum*. . . . . Bristly lady's thumb
Bristly lady's-thumb, a native of Asia, was discovered in North America in 1910 near Philadelphia, Pennsylvania, and by 1918, it had spread northeast into southern New York (Kochman 1991). The first sighting of it in the

Hudson River basin was recorded near Troy, Rensselaer County, in 1936 (*H.D. House 24129* NYS). *Polygonum cespitosum* var. *longisetum* grows along roadsides, waste places, shores, and other damp spots (Fernald 1950). Its mechanism of introduction remains unknown (Mills et al. 1993).

Polygonaceae
Polygonum convolvulus . . . . Black bindweed
Polygonum convolvulus has been naturalized from Europe and grows in cultivated areas and waste grounds (Britton and Brown 1896; Fernald 1950; Gleason and Cronquist 1991; Gray 1848; Torrey 1843). It was first recorded in the Hudson River basin in July 1869, in Poughkeepsie, Dutchess County (*W.R. Gerard s.n.* NYS). The mechanism by which it entered the Hudson River basin is unknown, although Brown (1879) found *Polygonum convolvulus* on dry ballast in New York City, possibly indicating its introduction via ships' dry ballast.

Polygonaceae
Polygonum hydropiper . . . . Water-pepper
Water-pepper was first recorded in the Hudson River basin in September 1926, at Guilderland Center, Albany County (*H.D. House 13327* NYS). This Eurasian native grows in low, moist ground (Britton and Brown 1896; Fernald 1950; Gleason and Cronquist 1991; Gray 1848) and was used by early settlers as a yellow dye for woolen clothing (Torrey 1843). It is not known how *Polygonum hydropiper* was introduced into the Hudson River basin.

Polygonaceae
Polygonum lapathifolium . . Dock-leaved smartweed
This European plant prefers swampy thickets, shores, damp clearings, cultivated fields and waste places (Britton and Brown 1896; Fernald 1950; Gleason and Cronquist 1991). *Polygonum lapathifolium* was first recorded in the Hudson River basin in August 1898, in Highland, Ulster County (*C.J. Elting 2288* NYS). The mechanism by which it entered the Hudson River basin is unknown. However, Brown (1879) found *Polygonum lapathifolium* on dry ballast in New York City, possibly indicating its introduction was via ships' dry ballast.

Polygonaceae
Polygonum persicaria . . . . Lady's thumb
Lady's thumb, a native of Europe, is common in damp clearings, gardens, along roadsides, shores, and waste places (Britton and Brown 1896; Fernald 1950; Gleason and Cronquist 1991; Gray 1848; Torrey 1843). It was not actually recorded in the Hudson River basin until August 1892, in Highland, Ulster County (*C.J. Elting 2295* NYS), although by 1843 it was considered naturalized throughout much of New York (Torrey 1843). It is not known how *Polygonum persicaria* was introduced into the Hudson River basin.

Polygonaceae
Rumex crispus . . . . Curly dock
Native to Europe, *Rumex crispus* is common and troublesome about houses, pastures, and meadows, and inhabits both cultivated and waste grounds (Britton and Brown 1896; Fernald 1950; Gleason and Cronquist 1991; Gray 1848; Torrey 1843). Curly dock was first recorded in the Hudson River basin in July 1924, near Green Island, Albany County (*H.D. House 10387* NYS). The mechanism by which it entered the Hudson River basin is unknown, although it probably escaped from cultivation.

Polygonaceae
Rumex obtusifolius . . . . Bitter dock
Bitter dock is an European weed that is now rather common in moist meadows, pastures, along roadsides, and waste places in the United States (Britton and Brown 1896; Fernald 1950; Gleason and Cronquist 1991; Gray 1848; Torrey

1843). It was first recorded in the Hudson River basin in September 1927, in Guilderland Center, Albany County (*H.D. House 24672* NYS). It is not known how *Rumex obtusifolius* arrived in the Hudson River basin.

Salicaceae

Salix alba . . . . White willow

Salix alba is an European tree that was known to have escaped from cultivation along stream banks and lake shores by the early to mid-1800s (Britton and Brown 1896; Fernald 1950; Gleason and Cronquist 1991; Gray 1848; Mitchell 1986; Nuttall 1818). The tree was widely imported into North America for ornamental purposes and its wood, bark, and leaves had medicinal and practical uses (Usher 1974). *Salix alba* was first recorded in 1851, near Amenia, Dutchess County (*A. Winchell observation* NYS).

Salicaceae

Salix fragilis . . . . Crack willow

This European willow was also known to escape from cultivation to roadsides and the borders of woods and was considered naturalized by the mid-1800s (Fernald 1950; Gleason and Cronquist 1991; Gray 1848; Nuttall 1818). The first record of *Salix fragilis* was made in the Hudson River basin in 1898, when it was collected between Yonkers and Harriman, Westchester County (*Bicknell s.n.* NYS). Twigs of the crack willow can break away, take root, and grow into other trees (Britton and Brown 1896). Gray (1848) indicated that crack willow was also cultivated for use in basket work.

Salicaceae

Salix purpurea . . . . Purple osier

Originally introduced from Europe for basket making (Fernald 1950), *Salix purpurea* escaped from cultivation along low ground and was naturalized by the early to mid-1800s (Gray 1848; Gleason and Cronquist 1991; Nuttall 1818;). *Salix purpurea* was first recorded in the Hudson River basin in 1920, in Albany, Albany County (*H.D. House 6975* NYS).

Brassicaceae

Cardamine pratensis . . . . Cuckoo-flower

Cardamine pratensis, a Eurasian native, grows in wet places such as swamps and bogs (Gray 1848; Torrey 1843) as well as in also lawns and meadows (Britton and Brown 1896; Fernald 1950, Gleason and Cronquist 1991). It was first recorded in the Hudson River basin in 1864 at Sander's Lake, Schenectady County (Paige 1864). It is not known how cuckoo-flower was introduced into the Hudson River basin.

Brassicaceae

Rorippa nasturtium-aquaticum . . . . Water-cress

Introduced from Europe (Fernald 1950), true water-cress was first recorded in the Hudson River basin in June 1893, in Highland, Ulster County (*C.J. Elting 193* NYS). It was widely cultivated as a salad plant (Britton and Brown 1896) and has escaped from cultivation (Gray 1848) into cool, clear waters of brooks and streams (Britton and Brown 1896; Fernald 1950; Gleason and Cronquist 1991).

Brassicaceae

Rorippa palustris
var. *palustris* . . . . Common yellow-cress

Rorippa palustris var. *palustris* is common along stream banks and rivers and other wet places in general (Britton and Brown 1896; Fernald 1950; Gleason and Cronquist 1991; Gray 1848; Torrey 1843). The first record of this Eurasian aquatic plant (Fernald 1950) in the Hudson River basin was made in 1875, when it was collected near Pine Plains, Dutchess County (*L.H. Hoysradt s.n.* HHH). Brown (1879) found common yellow-cress on a dry ballast dumping ground in New York City, possibly indicating its introduction into the Hudson River basin with the solid ballast of trans-Atlantic ships.

Brassicaceae

Rorippa sylvestris . . . . .Creeping yellow-cress

Rorippa sylvestris was first recorded in the Hudson River basin in 1896, in New Paltz, Ulster County (*C.J. Elting 198* NYS). This European plant grows in low grounds, waste places, meadows, shores, and along roadsides (Britton and Brown 1896; Fernald 1950; Gleason and Cronquist 1991). Brown (1879) also found creeping yellow-cress on dry ballast in New York City which may have been the entry mechanism for it in the Hudson River basin.

Primulaceae

Lysimachia nummularia . . . . .Moneywort

George W. Clinton, former member of the New York State Board of Regents, was the first to record the presence of moneywort in the Hudson River basin when he collected it in 1867 at Saratoga, Saratoga County (*G.W. Clinton s.n.* BUF). Gleason and Cronquist (1991) and Mitchell (1986) have reported this European plant as escaping from cultivation in gardens to damp roadsides, grasslands, and shores throughout the northern United States.

Primulaceae

Lysimachia vulgaris . . . . .Garden-loosestrife

Introduced from Europe (Fernald 1950), *Lysimachia vulgaris* was first recorded in the Hudson River basin in 1893, in Ohioville, Ulster County (*C.J. Elting 1711* NYS). Garden-loosestrife has been known to escape from cultivation to fields, roadsides, and swampy thickets (Gleason and Cronquist 1963; Mitchell 1986).

Haloragaceae

Myriophyllum spicatum . . .European water-milfoil

European water-milfoil closely resembles the native North American, *Myriophyllum exalbescens,* and it was often misidentified in early reports of North American flora. The first record of *Myriophyllum spicatum* in North America was made in the Potomac River in 1881 (Reed 1977). By 1970, it had expanded its range into the tidal Hudson River at Constitution Island, Putnam County (Reed 1970; *S.J. Smith observation* NYS) and was well established in Tivoli Bay in 1971 (Kiviat, personal communication). This plant grows in dense stands and can outcompete native plants through shading. It entangles boat propellers, clogs water intake devices, and its decaying mats can degrade the quality of beaches and add to maintenance costs (Aiken et al. 1979). European water-milfoil is a common aquarium plant and has spread as a result of unintentional aquarium release (Reed 1977) and by means of boats and boat trailers (Dove and Malcolm 1980; Scales and Bryan 1979).

Haloragaceae

Myriophyllum heterophyllum

The first recorded sighting of *Myriophyllum heterophyllum* in New York State was in 1953 from the Ramapo River drainage in Kanawauke Lake, Orange County (*J.C. Orth 290* TRLSD). It was later found at the West Point United States Military Academy Reservation (*R.S. Mitchell 7882* NYS) and Lake Tiorati, Orange County (*G.C. Tucker 9129* NYS).

Lythraceae

Lythrum salicaria . . . . .Purple loosestrife

Considered native to Eurasia, purple loosestrife has been known from North America since the early nineteenth century, and specimens were first obtained from the Hudson River basin in 1867 (Stuckey 1980). By the late 1800s, purple loosestrife was common throughout much of New York (Malecki 1987) and now occurs in nearly all parts of the United States, especially within the glaciated region of northeastern North America (Stuckey 1980). Its mode of transport is uncertain, although early records from eastern port cities suggest that the seeds may have been present in the discarded solid ballast of ships

(Rawinski 1982; Stuckey 1980). Stuckey (1980) also states that seeds were spread in waste from woolen mills and that the plants escaped from cultivation in gardens and nurseries.

Trapaceae

Trapa natans . . . . .Water chestnut
Native to temperate southern Europe and Asia (Muenscher 1944), water chestnut was introduced into North America in Middlesex County, Massachusetts in 1875 (Countryman 1970). It was first introduced into the Hudson River basin in Collins Lake near Scotia in 1884 (Wibbe 1886). By 1934, it had spread east from Collins Lake through its outflow into the Mohawk River and was abundant along the shores to Lock 6 near Waterford (Muenscher 1935a). The initial introduction mechanism of water chestnut is unknown, although Muenscher (1935b) reported that its fruits were being sold at fairs in western New York and the seeds were being sold by an aquarium dealer. It is now present in the lower Mohawk River and south through the Hudson River to Iona Island, Rockland County (Kiviat 1987). Water chestnut appears to shade out and reduce populations of submerged aquatic plants (e.g., *Vallisneria americana*) that may be important to waterfowl. Its spiked nuts are a nuisance to bathers and fishers, and the plant is a major impediment to boaters (Kiviat 1987).

Onagraceae

Epilobium hirsutum . . . . .Hairy willow-herb
Naturalized from Europe (Fernald 1950), *Epilobium hirsutum* was first recorded in the Hudson River basin in September 1866, in Albany, Albany County (*C.H. Peck s.n.* NYS). Found on ballast dumping grounds in New York City (Britton and Brown 1896; Brown 1878), hairy willow-herb has since spread into waste places, thickets, meadows, and along roadsides (Fernald 1950; Gleason and Cronquist 1991; Stuckey 1970).

Solanaceae

Solanum dulcamara . . . . .Bittersweet
This European native was initially imported and used by early European settlers as a remedy for rheumatism and scurvy (Torrey 1843). Most likely by escaping from cultivation, *Solanum dulcamara* was distributed around dwellings and moist stream and river banks throughout New York by 1843 (Gray 1848; Torrey 1843). The root, when chewed, is first bitter and then sweet-hence, the common name (Torrey 1843).

Menyanthaceae

Nymphoides peltata . . . . .Yellow floating-heart
In a letter to Muenscher (1933), former New York State Botanist Homer D. House wrote, "*Nymphoides peltatum* was found in the Hudson River at Waterford, September 2, 1929, by W. H. Barker. Since then he has seen it abundantly some distance upstream from there." *Nymphoides peltatum* is native to Europe and has spread from cultivation in eastern New York (Fassett 1957; Fernald 1950; Gleason and Cronquist 1991).

Boraginaceae

Myosotis discolor
(= M. versicolor) . .Yellow and blue scorpion-grass
This European native (Fernald 1950) was first collected in the Hudson River basin in 1882, in Riverdale, Bronx County (Bicknell 7328 NY). *Myosotis discolor* grows in fields, on hillsides, and along roads (Britton and Brown 1896; Fernald 1950; Gleason and Cronquist 1991; Gray 1848). The introduction mechanism for yellow and blue scorpion-grass is not known.

Boraginaceae

Myosotis scorpioides . . . . . . .Water scorpion-grass
Myosotis scorpioides was first recorded in the Hudson River basin in 1868, in Poughkeepsie, Dutchess County (*W.R. Gerard s.n.* NYS). Native to Europe, water scorpion-grass prefers shallow, wet habitats such as brooks and marshes

(Fernald 1950; Gleason and Cronquist 1991). Britton and Brown (1896) indicated that *Myosotis scorpioides* has escaped from cultivation and thereby extended its range in northeastern North America.

Lamiaceae

Lycopus europaeus . . .European water-horehound

The first two North American records for European water-horehound are from Norfolk, Virginia around 1860, and from solid ballast dumping grounds in the Delaware River in New Jersey in 1867 (Stuckey and Phillips 1970). Although this plant was recorded from ballast grounds in Philadelphia in 1876 (Martindale 1876) and in New York City in 1878 (Brown 1879), it was not recorded for other areas in the Hudson River basin until June 1966, when it was collected in Albany, Albany County (*S.J. Smith 40356* NYS). Fernald (1950) indicated that *Lycopus europaeus* occurs in waste places, ditches, and in the neighborhood of settlements, suggesting it has escaped from cultivation as well.

Lamiaceae

Mentha arvensis x spicata = M. x gentilis . . .Red mint
Mentha aquatica x spicata = M. x piperita .Peppermint
Mentha spicata . . . . .Spearmint

The mint genera is complex, and many species are the result of hybrids between and among those introduced from Europe and the only native North American mint, *Mentha arvensis*. The taxonomy of *Mentha* has changed considerably over the past 150 years, and the documentation of its arrival and dispersal would require extensive study of herbarium specimens, which is beyond the scope of this study. All the above mints have escaped from cultivation (Gleason and Cronquist 1991) and are considered naturalized (Mitchell 1986). By the mid-nineteenth century, peppermint was growing freely along the shores of the Hudson River (Torrey 1843). Peppermint is a sterile hybrid cultivated for its oil which is used as a flavoring (Usher 1974). Spearmint is a European native that is grown for its oil which is used medicinally and for flavoring (Usher 1974). By the mid-1800s, Gray (1848) and Torrey (1843) had found this plant common along stream banks and in wet meadows.

Lamiaceae

Stachys palustris var. *palustris* . . . . . .Hedge-nettle

Introduced from Europe (Fernald 1950), the first record of *Stachys palustris* var. *palustris* in the Hudson River basin was 1867 where it was collected near Poughkeepsie, Dutchess County (*W.R. Gerard s.n.* VAS). In the mid-1800s, hedge-nettle was more common in the western part of the state (Gray 1848; Torrey 1843), but it now occurs in the Hudson River basin near the borders of woods, in meadows, and along river banks and roadsides (Britton and Brown 1896; Fernald 1950; Gleason and Cronquist 1991). Brown (1879) and Martindale (1876) found *Stachys palustris* var. *palustris* on dry ballast in Philadelphia and New York City, respectively, suggesting that it was introduced with the ballast of trans-Atlantic ships.

Scrophulariaceae

Veronica serpyllifolia
var. *serpyllifolia* . . . . . . .Thyme-leaved speedwell

Veronica serpyllifolia var. *serpyllifolia* was first recorded in the Hudson River basin in 1836, in Troy, Rensselaer County (Wright and Hall, Troy). Native to Europe, thyme-leaved speedwell grows in a variety of habitats including fields, pastures, meadows, along roadsides, thickets, damp open woods, and lawns (Britton and Brown 1896; Fernald 1950; Gleason and Cronquist 1991; Gray 1848; Torrey 1843). The entry mechanism of *Veronica serpyllifolia* var. *serpyllifolia* is not known.

Rubiaceae

Galium aparine . . . . .Cleavers

This Eurasian plant grows in various habitats such as shady thickets, the margins of woods and fences, seashores, and waste grounds (Britton and Brown 1896; Fernald 1950; Gleason and Cronquist 1991; Gray 1848; Torrey 1843). *Galium aparine* was first recorded in the Hudson River basin in 1927, in Glenmont, Albany County (*C.H. Peck s.n.* NYS). It is not known how cleavers was introduced into the Hudson River basin.

Asteraceae

Bidens tripartita . . . . .Beggar-ticks

This European native was found on dry ballast dumping grounds in New York City in 1879 (Brown 1879), indicating that it probably arrived with the solid ballast of trans-Atlantic ships. Beggar-ticks was first recorded in the Hudson River basin from Glenmont, Albany County, in 1924 (*H.D. House 10640* NYS). *Bidens tripartita* is quite common in low, wet areas such as swamps (Britton and Brown 1896; Fernald 1950; Gleason and Cronquist 1991; Gray 1848; Torrey 1843).

Asteraceae

Boltonia asteroides var. *recognita* . . . . Boltonia

Boltonia asteroides var. *recognita* is native to prairies and streams of the central plains and has escaped from cultivation into the northeastern United States (Fernald 1950; Gleason and Cronquist 1991). Boltonia was first recorded in the Hudson River basin in 1929, near Brewster, Putnam County (*Mrs. R.F. Hodge* NYS).

Asteraceae

Helenium flexuosum
(= H. nudiflorum) . . . . .Southern sneezeweed

Native to the southeastern United States, sneezeweed has made its way northward and was first recorded in the Hudson River basin in 1904, in Jonesville, Saratoga County (*J.S. Cotton s.n.* NYS). *Helenium flexuosum* grows in the damp soil of fields, meadows, along roadsides, and waste grounds (Britton and Brown 1896, Fernald 1950, Gleason and Cronquist 1991). The entry mechanism of sneezeweed is not known.

Asteraceae

Sonchus arvensis . . . . .Perennial-sow-thistle

Native to Europe, Torrey (1843) found *Sonchus arvensis* near the quarantine area on Staten Island, possibly indicating its introduction with livestock as food or bedding. Brown (1878) also found it on dry ballast dumping grounds in New York City which could mean it was introduced with dry ballast as well. *Sonchus arvensis* was first recorded in the Hudson River basin in 1868, near Poughkeepsie, Dutchess County (*W.R. Gerard s.n.* VAS). Perennial-sow-thistle grows in low ground along salt marshes, shores, streams, fields, roadsides, and waste places (Britton and Brown 1896; Fernald 1950; Gleason and Cronquist 1991).

Butomaceae

Butomus umbellatus . . . . .Flowering rush

This Eurasian marsh and shore species was first collected in North America in 1905 at La Prairie, Quebec, a town across the St. Lawrence River from Montreal (Core 1941). The plant quickly spread, often as a result of deliberate introductions, and became established along a large part of the St. Lawrence River system, the Great Lakes, and northern Lake Champlain (Fernald 1950; Gaiser 1949; Gleason and Cronquist 1991). It was first recorded in the Hudson River basin in October 1965 near Schaghticoke, Rensselaer County (*S.J. Smith s.n.* NYS). The initial introduction of flowering rush into North America was probably the result of its release from a ship in Montreal, in the days when Montreal was a port where cargo was transferred from ocean-going ships, to lake ships until the canal system was expanded to accommodate larger vessels (Mills et al. 1993). It is not known how *Butomus umbellatus* arrived in the Hudson River basin, although it may have made its way through the Champlain Canal from Lake Champlain or it could have been released deliberately (Gaiser 1949).

Potamogetonaceae

Potamogeton crispus . . . . .Curly pondweed
This European plant can grow in fresh, brackish, or saltwater and will tolerate muddy or calcareous ponds and streams (Britton and Brown 1896; Fernald 1950; Gleason and Cronquist 1991). It was introduced into eastern North America and the Hudson basin sometime prior to 1814 (Muenscher 1944). Stuckey (1979) presented several incidents of *Potamogeton crispus* introductions associated with fish-stocking activities and also indicated that it has been planted as a source of waterfowl food and habitat.

Najadaceae

Najas minor . . . . .Eutrophic water-nymph
Najas minor is an Old World species that was discovered in the Hudson River basin in 1934 near Waterford at the mouth of the Mohawk River (Muenscher 1935a). It thrives in eutrophic and slow-moving waters and will tolerate mild pollution (Merilainen 1968). The means by which it initially arrived in North America is not known, although its transport within the state of New York was probably facilitated by waterfowl and shipping activity in the Erie Canal (Chase 1947). Fernald (1950) and Gleason and Cronquist (1991) noted *Najas minor* as "locally" established in the Hudson River valley, and this supports the notion of a direct inoculation into the basin (e.g., ballast discard at the port of Albany) as opposed to a more gradual spreading (e.g., as by escape from cultivation).

Juncaceae

Juncus inflexus . . . . .Rush
The first record of *Juncus inflexus* in the Hudson River basin was reported by Clarke and House (1921) after they found it well established in 1917 near Sangerfield and Waterville. This European native (Fernald 1950) had probably been introduced several years earlier, and although the area in which it was found had never previously been cultivated, it had been used for pasturage (Clarke and House 1921). Gleason and Cronquist (1991) noted *Juncus inflexus* from meadows, pastures, and damp roadsides, and thus it may have escaped from cultivation.

Poaceae

Agrostis gigantea
(= *A. stolonifera* var. *major*) . . . . .Redtop
Agrostis gigantea is native to Europe (Fernald 1950) and was first recorded near Lower Jay, Albany County, in 1924 (*H.D. House 10300* NYS). Redtop has escaped from cultivation into meadows, shores, coastal marshes, and other moist places (Gleason and Cronquist 1991), indicating the means by which it most likely spread into the Hudson River basin.

Poaceae

Alopecurus geniculatus . . . . Marsh-foxtail
Naturalized from Europe, *Alopecurus geniculatus* grows in wet meadows, ditches, ponds, and clearings (Britton and Brown 1896; Fernald 1950; Gleason and Cronquist 1991; Gray 1848; Torrey 1843). Marsh-foxtail was first recorded in the Hudson River basin in 1880 from dry ballast in New York City (*A. Brown s.n.* BH, Brown 1881), and therefore, it probably arrived on the solid ballast of trans-Atlantic ships.

Poaceae

Calamagrostis epigejos . . . . .Feathertop
Feathertop grows in sandy woods, salt marshes, fields, and waste places (Fernald 1950; Gleason

and Cronquist 1991). This European grass (Fernald 1950) was first recorded near Crescent, Saratoga County, in 1930 (*E.G. Whitney 1352* NYS). The entry mechanism of *Calamagrostis epigejos* remains unknown.

Poaceae
Echinochloa crusgalli . . . . .Barnyard-grass
Naturalized in North America from Eurasia, barnyard-grass has been cultivated for livestock forage (Dore and MacNeill 1980) and is common as a weed of cultivated land and waste ground (Britton and Brown 1896; Fernald 1950; Gleason and Cronquist 1991; Gray 1848; Torrey 1843). The first record of *Echinochloa crusgalli* in the Hudson River basin was made in August 1894, when it was collected near Highland, Ulster County (*C.J. Elting 3413* NYS). Its transport vector is not known, although it might have been intentionally released for agricultural purposes followed by its escape from cultivation.

Poaceae
Microstegium vimineum . . . . . .Japanese stilt-grass
Japanese stilt-grass is an Asian native that has invaded roadsides and open-forests in the Hudson River basin (Tucker 1994). Though the first sighting of *Microstegium vimineum* may have been in the mid-1970s, it was first collected in 1987 at Iona Island (Zaremba). Japanese stilt-grass has spread throughout the Hudson River basin as far north as Athens, Greene County, and is abundant at Harriman and Bear Mountain State Parks (Tucker 1994). The entry mechanism is not known.

Poaceae
Phragmites australis . . . . .Common reed
According to Mitchell (1986), this plant has native and exotic populations in New York. Fernald (1950) considered it native to Eurasia, although he did admit that its distribution is nearly cosmopolitan when considering all its varieties. The earliest record of *Phragmites australis* in the Hudson River basin was made in August 1844, when it was collected near Pine Plains and Schenectady, Schenectady County (*Pearson s.n.* NYS). It was most likely introduced some time before this because Torrey (1843) indicated that it was found in most parts of the world. Burk (1877) and Martindale (1876) found *Phragmites australis* on dry ballast in Philadelphia, suggesting this plant may have arrived in ship dry ballast. This plant prefers to grow in ditches and near the edges of ponds, swamps, and other wet places (Britton and Brown 1896; Gleason and Cronquist 1991; Gray 1848; Torrey 1843).

Poaceae
Poa annua . . . . .Speargrass
Low speargrass, an Eurasian native, grows in cultivated and waste grounds, along roadsides, clearings and garden paths (Britton and Brown 1896; Fernald 1950; Gleason and Cronquist 1991; Gray 1848; Torrey 1843). Based on these habitats, *Poa annua* probably escaped from cultivation and subsequently spread to other areas. It was first recorded in the Hudson River basin in August 1921, at Napanoch, Ulster County (*H.D. House* NYS), although it was probably introduced long before then.

Poaceae
Poa nemoralis . . . . .Wood bluegrass
Native to Eurasia (Fernald 1950), wood bluegrass was first recorded in the Hudson River basin in 1892 at Davenport Place, Bronx County (*Bicknell 11165* NYS). *Poa nemoralis* grows in a variety of habitats including thickets, moist meadows, open woods, shaded hills, rocky slopes, open sands, and shores (Fernald 1950; Gleason and Cronquist 1991; Gray 1848). It is not known how wood bluegrass was introduced into the Hudson River basin, although it probably escaped from cultivation.

Poaceae

Poa trivialis . . . . Rough bluegrass

This European grass (Fernald 1950) was commonly used as pasturage for livestock in the 1800s (Britton and Brown 1896; Gray 1848; Torrey 1843). It was present on Manhattan Island as early as 1811 (*Le Conte 1811*) and was also found on ballast waste there in 1879 (Brown 1880). The details of its introduction are unclear, although it was probably brought to North America with livestock as bedding or as a source of livestock food (Mills et al. 1993).

Iridaceae

Iris pseudacorus . . . . Yellow flag

This European native was grown as a common ornamental flower, and it was used for treating ailments such as toothaches, tanning leather and as a dye (Usher 1974). Yellow flag has subsequently escaped from cultivation into marshes, meadows, and brooksides (Cody 1961; Gleason and Cronquist 1991). It was first recorded in the Hudson River basin in 1868, near Poughkeepsie, Dutchess County (*W.R. Gerard s.n.* NYS) and is now frequent along the Hudson River in northern Dutchess County (Kiviat, personal communication).

Discussion

The Hudson River basin is a young biological system in its most modern manifestation. Wisconsin Ice Age glaciers covered the basin 18,000 years before present (BP), and most of the basin was again ice-free 12,000 BP (Thompson 1966). The Mohawk River Valley, formed 13,000 BP, provided drainage from the Great Lakes until a lower drainage at Niagara Falls was formed 1,000 years later (Thompson 1966). The indigenous freshwater biota of the modern Hudson River basin is thus derived from adjacent geographic regions and from lower aboriginal freshwater systems (now drowned on the continental shelf due to sea level rise), whose biota progressed inland as the seas rose and the glaciers receded. In many such late Tertiary glaciated environments—terrestrial, freshwater, and marine—the relatively low-diversity biotas that resulted from these recent natural colonization events in the last few millennia have been readily susceptible to modern human-mediated invasions in the last few centuries. The Hudson River is no exception.

Source Regions and Mechanisms

Since the early 1800s, at least 113 exotic species have entered the Hudson River basin (Figure 3). Aquatic plants (47%) and fish (28%) comprise the majority of these species. Eliminating the poorly known invertebrate group, mollusks and crayfish represent the remaining 21% of introduced species.

A variety of entry mechanisms, including unintentional release, shipping activities, canals, and intentional release, have played important roles in transporting exotic species into the Hudson River basin (Figure 4). Unintentional release (25%) was clearly the most important transport mechanism, and within this category,

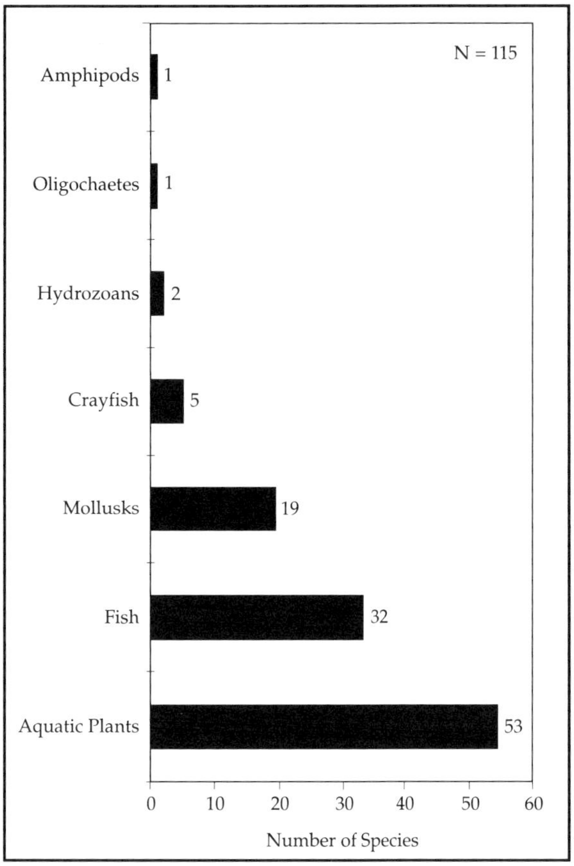

Figure 3. Introduced aquatic species in the Hudson River sorted by taxonomic group.

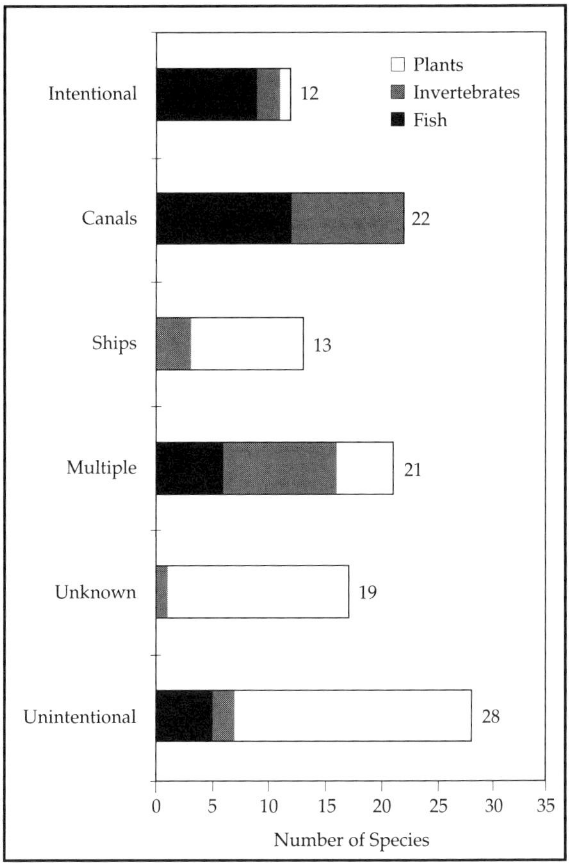

Figure 4. Entry mechanisms of exotic species in the Hudson River basin sorted by taxonomic group.

84% of the releases have resulted from the escape of plants from cultivation. Nearly 12% of the exotic species in the Hudson River basin have been associated with ships, of which 76% have been plants transported to North America in ship solid ballast. Canals were associated with 19% of overall introductions into the Hudson River basin, and most of these introductions were fish and invertebrates. The intentional release of organisms into the Hudson River basin was the least common vector, accounting for only 12% of total introductions, but 28% of fish introductions. For species that were transported to the basin by more than one mechanism (considered the ìmultipleî mechanism category), the most common vectors were found to be unintentional (35%), intentional (23%), shipping (16%), and canals (26%). For 15% of the exotic species, no records were available on how these organisms entered the Hudson River basin (i.e., "unknown" category).

The overall number of non-indigenous fish, plants, and invertebrates established in the Hudson River basin appears to have increased during the last 30 years due primarily to an increase in the number of fishes and invertebrates (Figure 5). Introductions associated with shipping and intentional releases have decreased in the last 60 years (Figure 6). The introduction of fishes and invertebrates entering the basin through canals has increased since 1930, while the number of unintentional releases has decreased since 1900, corresponding to a decline in the number of introduced plant species. This specific decline in unintentional releases was likely linked with a

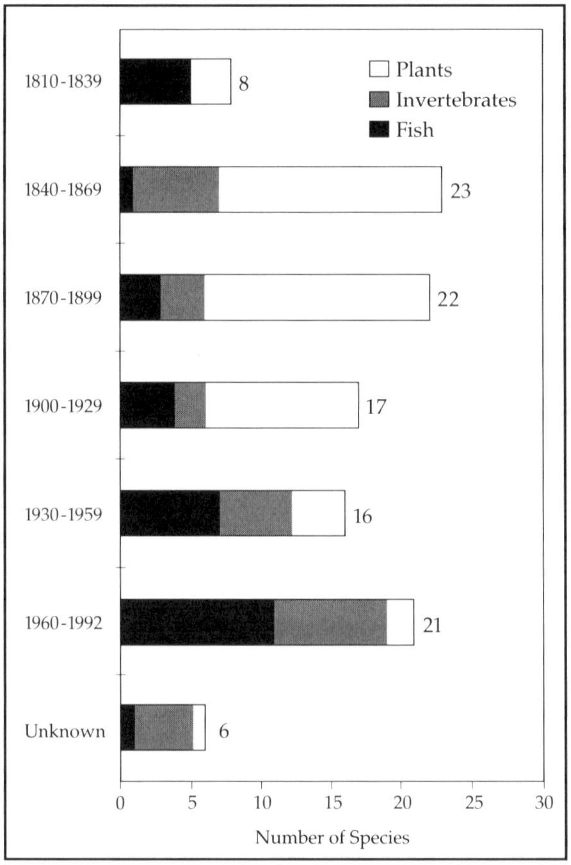

Figure 5. Exotic species in the Hudson River basin sorted by taxonomic group, 1810-1992.

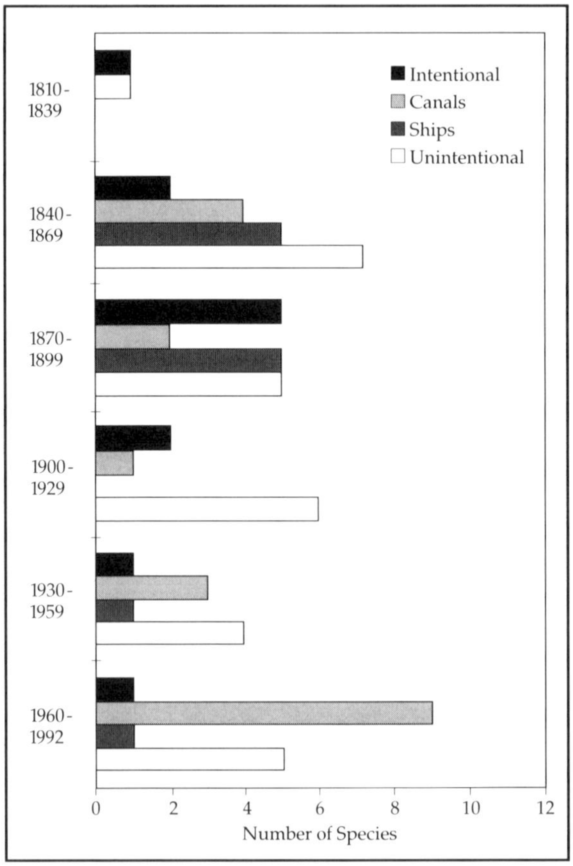

Figure 6. Entry mechanisms for exotic species in the Hudson River basin, 1810-1992.

historic decline in plant species escaping from cultivation and solid ballast.

Similar to those in the Great Lakes (Mills et al. 1993), most exotic species in the Hudson River basin are native to Eurasia (56%) (Figure 7). Despite the large number of exotic species native to Europe and Eurasia, however, some species have been introduced to the Hudson River basin from outside their native range. For example, 30% of all non-indigenous species in the Hudson River originated from the Interior Basin of the United States. Several species of aquatic invertebrates from the southern United States have invaded the Hudson River; these include the gammarid amphipod *Gammarus daiberi*, and the bivalves *Mytilopsis leucophaeata* and *Rangia cuneata*. Aquatic organisms have also been transported to the Hudson River from different parts of North America by ballast water. It is probable that the recent appearance of the southern clam *Rangia* is due to ballast water release in the mid- to late 1980s.

Extent of Underestimation of the Introduced Biota

As noted under "Methods," severe limitations attend the recognition of non-native invertebrates other than mollusks and crayfish. Our inclusion of only one other crustacean (a gammarid amphipod), two hydroids, and one worm represents a gross underestimation of introduced biota in a

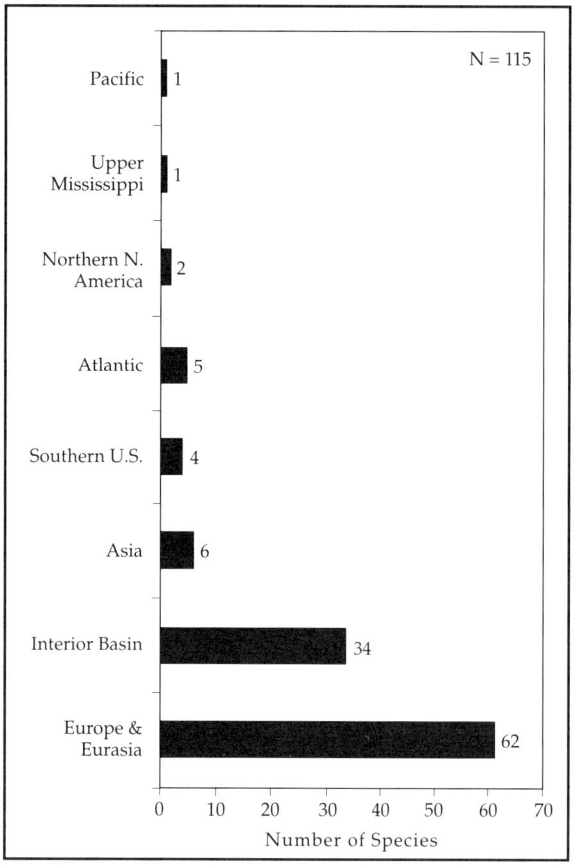

Figure 7. Geographic source of introduced species in the Hudson River basin.

region whose natural history has ostensibly been under study for almost 200 years. Indeed, such common freshwater exotics well known from eastern North America as the European weevil *Tanysphyrus lemnae* (which occurs virtually everywhere the duckweed *Lemna* occurs [Pennak 1978], the latter being common in the Hudson River), the European flatworm *Dugesia polychroa* (which occurs as close as Lake Champlain [Ball 1969]), the Asian oligochaete *Branchiura sowerbyi* (Brinkhurst 1965), and the Asian copepod *Argulus japonicus* (parasitic on goldfish [Cressey 1978], which occur in the Hudson), appear to remain unrecorded from the Hudson River system.

We emphasize that it would thus be erroneous to state that "113 species" have been introduced to the Hudson River basin, when there have been no analyses of algae, diatoms, protozoans, sponges, most hydrozoans, flatworms, rotifers, nematodes, most annelids, virtually none of the smaller crustacean groups, insects, bryozoans, and many other taxa. It is inconceivable that all the taxa in each of these groups are native in the Hudson River system. Indeed, if only two species among the sponges, flatworms, other hydrozoans, bryozoans, rotifers, and nematodes have been introduced, and only five species of each of the other taxa noted, then at least 40 additional species of non-native animals and plants may be in the river system. For most invertebrates, there are no lists of species present in the Hudson River, either because no one has looked (e.g., Ectoprocta) or because the species-level taxonomy has not been well worked out. Also, for most of these invertebrates, any survey work done is generally recent and spotty. The absence of extensive, older surveys greatly reduces our ability to recognize exotic species because we often recognize exotic species as (1) those that are widely distributed in their assumed native range, but patchy (especially near centers of human activity) in their introduced range, or (2) those not formerly found in the assumed introduced range or expanding within that range. The absence of historical data on Hudson River exotics has also limited our ability to trace exotics that may have become established and later disappeared. For example, Atlantic salmon (*Salmo salar*) was deliberately introduced in the Hudson River in the 1880s (Cheney 1887; Mather 1889b) with numerous collections of these fish reported in 1888 (Mather 1889b). Bean (1903) listed records of Atlantic salmon in the Hudson River around the turn of the twentieth century, but by 1936, this fish species had apparently disappeared (Greeley 1937). Finally, this review represents only a beginning to assessing exotics species in the Hudson River. The number of exotics will undoubtedly swell with a rigorous evaluation of specimens in herbaria and museums and the inclusion of terrestrial plants.

The Growing Problem of Exotics

A recent review of non-indigenous species in the United States (Office of Technology Assessment 1993) indicated that at least 4,500 species of foreign origin have become established over the last 100 years. These include several thousand plant and insect species and several hundred non-indigenous vertebrate, mollusk, fish, and plant pathogen species. The Office of Technology Assessment approximates that 2 to 8% of each group of organisms is non-indigenous to the United States; we can only suspect that this percentage applies to the Hudson River as well. In addition, the Office of Technology Assessment suggests that approximately 15% of all exotic species in the United States cause significant harm. This figure is comparable to findings in the Great Lakes which show that nearly 10% of all exotics have had significant impacts (Mills et al. 1993). Unfortunately, the number of exotic species that have had significant impacts on the Hudson River system has not been well documented, so we can only suspect that 10 to 15% of the total exotic species pool has had significant impacts here as well. Judging from other studies, such species as the zebra mussel, purple loosestrife, common carp, European water-milfoil, and water chestnut have had significant impacts on the Hudson River ecosystem. Losses from non-indigenous species can be economic as well as ecological; in the United States alone, 79 exotic species have produced documented economic losses of $97 billion from 1906 to 1991 (Office of Technology Assessment 1993). Studies examining the impact of zebra mussels in the Hudson River are currently underway, but information on the impacts of other exotics in this ecosystem is not readily available. Other impacts need to be addressed, including those related to the effects of introduced biota on native biological diversity and the effects of introduced biota on the economy of the Hudson Valley.

In the future, we can expect that the natural movement of non-indigenous species into the Hudson River will be less of a threat and that vectors associated with human activity and habitat alteration will continue to provide opportunities for the establishment of new species. The only exception is associated with global climatic change where species tolerant of higher temperatures could become established. Inland canals such as the Erie Canal will continue to allow the mixing of exotics between water bodies such as the Great Lakes, the St. Lawrence River, and Lake Champlain. We estimate that 32 exotic species (19%) have used canals to enter the Hudson. Of course, exotics in the Hudson can also use canals to invade other waters. For example, the sea lamprey, alewife, and white perch all have moved from the Champlain and Hudson drainage through the Erie Canal to the Great Lakes (Mills et al. 1993). Different groups of organisms arrive in the Hudson River basin by different pathways, but we can expect future unintentional releases associated with such activities as ship ballast water, aquarium trade, and the bait fish industry to provide the mechanisms for the establishment of exotics into the Hudson River ecosystem. While it would appear that the bulk of the exotics becoming established in the Hudson River basin did so around 1900, one should not assume that the rate of invasion by exotics has necessarily declined over the last 90 years. There are significant voids in the historical species database for the Hudson, especially for invertebrates, crustaceans, and microscopic organisms. Such gaps in our knowledge only raise doubts about whether species invasions are increasing or decreasing. In the Great Lakes where patterns of invasion are better documented, 30% of all exotics have become established in the last 30 years, a surge coinciding with the opening of the St. Lawrence Seaway in 1959. Finally, the continuing threat of biological invasions in the Hudson River ecosystem is real, although the timing and extent of the invasions are impossible to predict.

Literature Cited

Adams, A.G. 1981. The Hudson. A guidebook to the river. State University New York Press, Albany, NY.

Aiken, S.G., P.R. Newroth, and I. Wile. 1979. The biology of Canadian weeds 34. *Myriophyllum spicatum* (Linnaeus). Canadian Journal of Plant Sciences 59:201-215.

Aldrich, T.H. 1869. Partial list of shells found near Troy, New York. Annual Report of the New York State Cabinet of Natural History 22:17-24.

Anonymous. 1993. Rudd ruckus. Fisheries 18(2):36.

Bade, E. 1926. The central European bitterling found in the United States. Bulletin of the New York Zoological Society 1926(6):188, 205-206.

Baker, F.C. 1898. The molluscan fauna of western New York. Transactions of the Academy of Science of St. Louis 8:71-94.

Ball, I.R. 1969. *Dugesia lugubris* (Tricladida: Paludicola), a European immigrant into North American fresh waters. Journal of Fisheries Research Board of Canada 26(2):221-228.

Bath, D.C., A. Beebe, C.B. Dew, R.H. Reider, and J.H. Hecht. 1976. A list of common and scientific names of fishes collected from the Hudson River. 7 pp. in Hudson River ecology: fourth symposium on Hudson River ecology, held at Bear Mountain, New York, March 28-30, 1976, sponsored by the Hudson River Environmental Society, Inc.

Bean, T.H. 1897. Notes upon New York fishes received at the New York Aquarium, 1895-1897. Bulletin of the American Museum of Natural History 9:327-375.

Bean, T.H. 1903. Catalogue of the fishes of New York. New York State Museum Bulletin 60.

Beauchamp, W.M. 1880. *Bythinia tentaculata* (Linnaeus). American Naturalist 14:523.

Beebe, C.A., and I.R. Savidge. 1988. Historical perspective on fish species composition and distribution in the Hudson River estuary. American Fisheries Society Monograph 4:25-36.

Bousfield, E.L. 1969. New records of *Gammarus* (Crustacea: Amphipoda) from the Middle Atlantic region. Chesapeake Science 10:1-17.

Bousfield, E.L. 1973. Shallow-water gammaridean Amphipoda of New England. Cornell University Press, Ithaca, NY.

Bovbjerg, R.V. 1970. Ecological isolation and competitive exclusion in two crayfish. Ecology 51:225-236.

Brinkhurst, R.O. 1965. Studies on the North American aquatic Oligochaeta II: Tubificidae. Proceedings of the Academy of Natural Sciences of Philadelphia 117(4):117-172.

Brinkhurst, R.O. 1986. Guide to the freshwater aquatic microdrile oligochaetes of North America. Canadian Special Publication of Fisheries and Aquatic Sciences 84:1-259.

Brinkhurst, R.O., and B.G. Jamieson. 1971. Aquatic Oligochaeta of the world. Oliver and Boyd, Edinburgh.

Britton, N.L., and A. Brown. 1896. An illustrated flora of the Northern United States, Canada and the British possessions from Newfoundland to the parallel of the southern boundary of Virginia, and from the Atlantic Ocean westward to the 102d meridian. Charles Scribner's Sons, New York. 3 vol.

Brown, A. 1878. Plants introduced with ballast and on made land. Bulletin of the Torrey Botanical Club 5:255-258.

Brown, A. 1879. Ballast plants in New York City and its vicinity. Bulletin of the Torrey Botanical Club 6:353-360.

Brown, A. 1880. Ballast plants in and near New York City. Bulletin of the Torrey Botanical Club 7:122-126.

Brown, A. 1881. Ballast plants in and near New York City. Bulletin of the Torrey Botanical Club 8:141-142.

Bruce, W. 1907. The Hudson: Three Centuries of History, Romance, and Invention. Bryant Union County, NY.

Burch, J.B. 1975. Freshwater Sphaeriacean clams (Mollusca: Pelecypoda) of North America. Biota of Freshwater Ecosystems Identification Manual Number 3.

Burch, J.B. 1982. Freshwater snails (Mollusca: Gastropoda) of North America. United States Environmental Protection Agency Publication 600/3-82-026.

Burk, I. 1877. List of plants recently collected on ships ballast in the neighborhood of Philadelphia. Transactions of the Academy of Natural Sciences of Philadelphia 29:105-109.

Burkhead, N.M., and J.D. Williams. 1991. An intergeneric hybrid of a native minor, the golden shiner, and an exotic minnow, the rudd. Transactions of the American Fishery Society 120:781-795.

Bushnell, J.H., Jr., and T.W. Porter. 1967. The occurrence, habitat, and prey of *Craspedacusta sowerbyi* (particularly polyp stage) in Michigan. Transactions of the American Microscopy Society 86(1):22-27.

Capelli, D.M. 1982. Displacement of northern Wisconsin crayfish by *Orconectes rusticus* (Girard). Limnology and Oceanography 27(4):741-745.

Carlton, J.T. 1985. Transoceanic and interoceanic dispersal of coastal marine organisms: the biology of ballast water. Oceanography and Marine Biology, An Annual Review 23:313-371.

Carlton, J.T. 1989. Man's role in changing the face of the ocean: biological invasions and implications for conservation of near-shore environments. Conservation Biology 3:265-273.

Carlton, J.T. 1992a. Dispersal of living organisms into aquatic ecosystems as mediated by aquaculture and fisheries activities. Pp. 13-45 in A. Rosenfield and R. Mann (eds.). Dispersal of living organisms into aquatic ecosystems. Maryland Sea Grant Publication, College Park, MD.

Carlton, J.T. 1992b. Blue immigrants: the marine biology of maritime history. The Log (Mystic Seaport Museum, Mystic, CT) 44:31-36.

Carlton, J.T. 1992c. Introduced marine and estuarine mollusks of North America: an end-of-the-20th-century perspective. Journal of Shellfish Research 11:489-505.

Carlton, J.T. 1993. Dispersal mechanisms of the zebra mussel (*Dreissena polymorpha*). Pp. 677-697 in T.F. Nalepa and D.W. Schloesser (eds). Zebra Mussels: Biology, Impacts, and Control. CRC Press, Inc., Boca Raton, FL.

Carlton, J.T., and J. Geller. 1993. Ecological roulette: the global transport and invasion of nonindigenous marine organisms. Science 261:78-82.

Carlton, J.T., and J. Hodder. 1995. Coastal dispersal of marine organisms on a replica of a sixteenth century sailing vessel: implications for biogeography and gene flow. Marine Biology 121:721-730.

Chase, S.S. 1947. Preliminary studies in the genus *Najas* in the United States. Ph.D. dissertation, Cornell University, Ithaca, NY. 118 pp.

Cheney, A.N. 1887. Salmon in the Hudson River. Bulletin of the United States Fish Commission 6:351-352.

Cheney, A.N. 1897a. Black bass and their distribution in the waters of New York State. Annual Report of the Commission of Fisheries, Game and Forestry New York 2(1896):176-184.

Cheney, A.N. 1897b. The pike-perch. Annual Report of the Commission of Fisheries, Game and Forestry New York 2(1896):203-206.

Clarke, A.H. 1981. The freshwater mollusks of Canada. National Museums of Canada, Ottawa.

Clarke, R. 1852. Fishes of northern New York. Scientific American 7:182.

Clarke, J.M., and H.D. House. 1921. Report of the state botanist for 1921. Bulletin of the New York State Museum 243:5-102.

Clench, W.J. 1962. A catalogue of the Viviparidae of North America with notes on the distribution of *Viviparus georgianus* (Linnaeus). Occasional Papers on Mollusks 2:261-287.

Clench, W.J., and S.L.H. Fuller. 1965. The genus *Viviparus* (Viviparidae) in North America. Occasional Papers on Mollusks 2:385-412.

Cody, W.J. 1961. *Iris pseudacorus* (Linnaeus) escaped from cultivation in Canada. Canadian Field-Naturalist 75:139-142.

Countryman, W.D. 1970. The history, spread and present distribution of some immigrant aquatic weeds in New England. Hyacinth Control Journal 8:50-52.

Courtenay, W.R. Jr., and J.R. Stauffer, Jr. 1990. The introduced fish problem and the aquarium fish industry. Journal of the World Aquaculture Society 21(3):145-159.

Core, E.L. 1941. *Butomus umbellatus* in America. Ohio Journal of Science 41:79-85.

Crandall, M.E. 1977. Epibenthic invertebrates of Croton Bay in the Hudson River. New York Fish and Game Journal 24(2):178-186.

Cressey, R.F. 1978. Marine flora and fauna of the northeastern United States. Crustacea: Branchiura. NOAA Technical Report Circular 413:1-10.

Crocker, D.W. 1957. The crayfishes of New York State (Decapoda: Asticidae). Bulletin of the New York State Museum 355.

Crocker, D.W. 1979. The crayfishes of New England. Proceedings of the Biological Society of Washington 92:225-252.

Daniels, R.A. 1993. Creation of the cosmopolitan creek: A history of change in fish assemblages. Clearwater 23(1):20-24.

Deevey, E.S., and J.L. Brooks. 1943. *Craspedacusta* in open water, Lake Quassapaug, Connecticut. Ecology 24:266-267.

DeKay, J.E. 1842. Zoology of New York or the New York fauna. Part IV. Fishes. W. and A. White and J. Visscher, Albany, NY.

Dence, W.A. 1925. Bitter carp (*Rhodeus amarus*) from New York State waters. Copeia 142:33.

Dew, C.B. 1973. Comments on the recent incidence of the gizzard shad, *Dorosoma cepedianum*, in the lower Hudson River. 10 pp. in Hudson River ecology: third symposium on Hudson River ecology, held at Bear Mountain, New York, March 22-23, 1973, sponsored by the Hudson River Environmental Society, Inc.

Dore, W.G., and J. MacNeill. 1980. Grasses of Ontario. Research Branch, Agriculture Canada Monograph 26.

Dove, R., and B. Malcolm. 1980. The 1979 Aquatic Plant Quarantine Project. Studies on aquatic macrophytes 32. Water Investigations Branch, Ministry of Environment, British Columbia, Canada.

Drake, J.A., H.A. Mooney, F. di Castri, R.H. Groves, F.J. Kruger, M. Rejmanek, and M. Williamson (eds.). 1989. Ecology of biological invasions: a global perspective. SCOPE 37. John Wiley and Sons, New York.

Elton, C. 1958. The ecology of invasions by animals and plants. Methuen, London.

Fassett, N.C. 1957. A manual of aquatic plants. The University of Wisconsin Press, Madison, WI.

Faxon, W. 1898. Observations on the Asticidae in the United States National Museum and the Museum of Comparative Zoology, with descriptions of new species. Proceedings of the United States National Museum 20(1136):643-694.

Fernald, M.L. 1950. Gray's manual of botany. 8th edition. American Book Company, New York.

Forney, J. 1956. Raising crayfish for bait. New York State Conservationist 10(6):6-7.

Gaiser, L.O. 1949. Further distribution of *Butomus umbellatus* in the Great Lakes Region. Rhodora 51:385-390.

Garman, H. 1916. The sudden appearance of great numbers of fresh-water medusae in a Kentucky Creek. Science 44(1146):858-860.

George, C.J. 1981. The fishes of the Adirondack Park. Unpublished manuscript. 94 pp. Located in New York State Department of Environmental Conservation, Bureau of Fisheries, 50 Wolf Road, Albany, NY 12233.

George, C.J. 1983. Occurrence of the gizzard shad in the lower Mohawk Valley. New York Fish and Game Journal 30(1):113-114.

George, C.J. 1985. Occurrence of the silver lamprey in the stillwater sector of the Hudson River. New York Fish and Game Journal 32(1):95.

Gleason, H.A., and A. Cronquist. 1991. Manual of vascular plants of Northeastern United States and adjacent Canada. D. Van Nostrand Company, Inc., Princeton, NJ.

Godfrey, P.K. 1979. Aquatic and wetland plants of southeastern United States: monocotyledons. University of Georgia Press, Athens, GA.

Godfrey, P.K. 1981. Aquatic and wetland plants of southeastern United States: dicotyledons. University of Georgia Press, Athens, GA.

Goode, G.B. 1903. American fishes: a popular treatise upon the game and food fishes of North America, with especial reference to habits and methods of capture. L.C. Page and Company, Boston, MA.

Goodrich, C. 1942. The Pleuroceridae of the Atlantic coastal plain. Occasiónal Papers of the Museum of Zoology, University Michigan 456:1-6.

Gray, A. 1848. Manual of the botany of the northern United States. Metcalf and Company, Cambridge, MA.

Gray, A.F. 1883. *Bithynia tentaculata* (Linnaeus). American Naturalist 17:205.

Greeley, J.R. 1930. Fishes of the Lake Champlain watershed. Pp. 44-87 in A biological survey of the Champlain watershed. New York State Conservation Department, Supplemental to Nineteenth Annual Report (1929).

Greeley, J.R. 1935. Fishes of the watershed with annotated list. Pp. 63-101 in A biological survey of the Mohawk-Hudson watershed. New York State Conservation Department, Supplemental to Twenty-fourth Annual Report (1934).

Greeley, J.R. 1937. Fishes of the area with annotated list. Pp. 45-103 in A biological survey of the Lower Hudson watershed. New York State Conservation Department, Supplemental to Twenty-sixth Annual Report (1936).

Greeley, J.R., and S.C. Bishop. 1933. Fishes of the area with annotated list. Pp. 64-101 in A biological survey of the Upper Hudson watershed. New York State Conservation Department, Supplemental to Twenty-second Annual Report (1932).

Greenfield, D.W. 1974. An evaluation of the advisability of the release of the grass carp, *Ctenopharyngodon idella*, into the natural waters of the United States. Transactions of the Illinois State Academy of Science 66:47-53.

Hamilton, A., and R.E. Schmidt. 1992. Creature feature: mosquitofish in the Hudson River. The Tidal Exchange 3(2):7.

Hebert, P.D.N., B.W. Muncaster, and G.L. Mackie. 1989. Ecological and genetic studies on *Dreissena polymorpha* (Pallas): a new mollusc in the Great Lakes. Canadian Journal of Fisheries and Aquatic Sciences 46:1587-1591.

Hengeveld, R. (editor). 1989. Dynamics of biological invasions. Chapman and Hall, London.

Herrington, H.B. 1962. A revision of the Sphaeriidae of North America (Mollusca: Pelecypoda). Miscellaneous Publications of the Museum of Zoology, University of Michigan 118:1-75.

Hervey, G.F., and J. Hems. 1948. The goldfish. The Batchwork Press, London.

Heywood, V.H. 1989. Patterns, extents and modes of invasions by terrestrial plants. Pp. 31-55 in J.A. Drake, H.A. Mooney, F. di Castri, R.H. Groves, F.J. Kruger, M. Rejmanek, and M. Williamson (eds). Ecology of biological invasions: a global perspective. SCOPE 37. John Wiley and Sons, New York.

Hiltunen, J.K., and D.J. Klemm. 1980. A guide to the Naididae (Annelida: Clitellata: Oligochaeta) of North America. United States Environmental Protection Agency Publication 600/4-80/031.

Hopkins, S.H., and J.D. Andrews. 1970. *Rangia cuneata* on the East Coast: thousand mile range extension, or resurgence. Science 167:868-869.

House, H.D. 1924. Annotated list of the ferns and flowering plants of New York State. Bulletin of the New York State Museum 254.

House, H.D. 1942. Bibliography of the botany of New York State 1751-1940. Bulletin of the New York State Museum 328:1-174, 329:125-233.

Jacobson, M.K. 1953. *Congeria leucophaeata* (Conrad) in the Hudson River. Nautilus 66(4):125-128.

Jokinen, E.H. 1982. *Cipangopaludina chinensis* (Gastropoda: Viviparidae) in North America, review and update. Nautilus 96(3):89-95.

Jokinen, E.H. 1992. The freshwater snails (Mollusca: Gastropoda) of New York State. Bulletin of the New York State Museum 482.

Kammen, M.G. 1975. Colonial New York: a history. KTO Press, White Plains, NY.

Kat, P.W. 1986. Hybridization in a unionid suture zone. Malacologia 27:107-125.

Kiviat, E. 1987. Water chestnut (*Trapa natans*). Pp 31-38 in D.J. Decker and J.W. Enck (eds.). Exotic plants with identified detrimental impacts on wildlife habitats in New York State. Natural Resources Research and Extension Series Number 29.

Klemm, D.J. 1985. A guide to the freshwater Annelida (Polychaeta, Naidid and Tubificid Oligochaeta, and Hirudinea) of North America. Kendall/Hunt Publishing Company, Dubuque, IA.

Kochman, A. 1991. *Polygonum cespitosum*: an Asian invader in the wetland flora of the eastern United States. Ohio Journal of Science 91(2):28.

Kornberg, H., and M.H. Williamson (editors). 1986. Quantitative aspects of the ecology of biological invasions. Philosophical Transactions of the Royal Society of London (B, Biological Sciences) 314:501-746.

Kramp, P.L. 1950. Freshwater medusae in China. Proceedings of the Zoological Society of London 120:165-184.

Lachner, E.A., C.R. Robins, and W.R. Courtenay, Jr. 1970. Exotic fishes and other aquatic organisms introduced into North America. Smithsonian Contributions to Zoology 59:1-29.

Latchford, F.R. 1914. *Valvata piscinalis* in Canada. Nautilus 28:10.

Latchford, F.R. 1925. *Bithynia tentaculata* (Linnaeus). Canadian Field-Naturalist 39:41-43.

Latchford, F.R. 1930. Some introduced molluscs. Canadian Field-Naturalist 44:33-34.

Leach, J.H. 1993. Impacts of the zebra mussel (*Dreissena polymorpha*) on water quality and fish spawning reefs in western Lake Erie. Pp. 381-397 in T.F. Nalepa and D.W. Schloesser (eds.). Zebra mussels: biology, impacts, and control. Lewis Publishers, Boca Raton, FL.

Le Conte, J. 1811. Catalogus Plantarum. American Medical and Philosophical Register 2:134-142.

Lee, D.S., C.R. Gilbert, C.H. Hocutt, R.E. Jenkins, D.E. McAllister, and J.R. Stauffer, Jr. 1980. Atlas of North American freshwater fishes. North Carolina State Museum of Natural History, Raleigh, NC.

Leidy, J. 1870. On the hydroid *Cordylophora*. Proceedings of the Academy of Natural Sciences of Philadelphia 22:113.

Letson, E.J. 1905. Checklist of the Mollusca of New York. Bulletin of the New York State Museum 88:1-112.

Lewis, J. 1856. Shellbearing species of Mollusca observed in portions of Herkimer and Otsego counties, New York. Proceedings of the Boston Society of Natural History 6:2-4.

Lewis, J. 1860. Catalogue of the mollusks in the vicinity of Mohawk, New York. Proceedings of the Academy of Natural Sciences of Philadelphia 12:17-19.

Lewis, J. 1872. Shells of Herkimer and adjacent counties in the state of New York. Proceedings of the Academy of Natural Sciences of Philadelphia 24:97-107.

Lodge, D.M., T.K. Kratz, and G.M. Capelli. 1986. Long-term dynamics of three crayfish species in Trout Lake, Wisconsin. Canadian Journal of Fisheries and Aquatic Sciences 43:993-998.

Lodge, D.M., and J.G. Lorman. 1987. Reductions in submersed macrophyte biomass and species richness by the crayfish *Orconectes rusticus*. Canadian Journal of Fisheries and Aquatic Sciences 44:591-597.

Lorman, J.G., and J.J. Magnuson. 1978. The role of crayfish in aquatic ecosystems. Fisheries 3(6):8-10.

Lytle, C.F. 1960. A note on distribution patterns in *Craspedacusta*. Transactions of the American Microscopy Society 79(3):461-469.

MacCrimmon, H.R. 1971. World distribution of rainbow trout (*Salmo gairdneri*). Journal of Fisheries Research Board of Canada 28(5):663-704.

Malecki, R. 1987. Purple loosestrife (*Lythrum salicaria*). Pp. 39-45 in D.J. Decker and J.W. Enck (eds.). Exotic plants with identified detrimental impacts on wildlife habitats in New York State. Natural Resources Research and Extension Series Number 29.

Marelli, D.C., and S. Gray. 1985. Comments on the status of recent members of the genus *Mytilopsis* (Bivalvia: Dreissenidae). Malacological Review 18:122-177.

Marshall, W.B. 1895. List of shells inhabiting the vicinity of Albany and Troy, New York. 48th Annual Report of the New York State Museum:641-647.

Martindale, I.C. 1876. The introduction of foreign plants. Botanical Gazette 2:55-58.

Mather, F. 1889a. Brown trout in America. Bulletin of the United States Fish Commission 7:21-22.

Mather, F. 1889b. Report upon the results of salmon planting in the Hudson River. Bulletin of the United States Fish Commission 7(1887):409-424.

Mayden, R.L., and C.R. Gilbert. 1989. *Notropis ludibundus* Girard and *Notropis tristis* Girard replacement names for *Notropis stramineus* Cope and *Notropis topeka* Gilbert Teleostei Cypriniformes. Copeia 1989:1084-1089.

Merilainen, J. 1968. *Najas minor* All. in North America. Rhodora 70:161-175.

Mills, E.L., J.L. Forney, and K.J. Wagner. 1987. Fish predation and its cascading effect on the Oneida Lake food chain. Pp. 118-131 in W.C. Kerfoot and A. Sih (eds.). Predation: direct and indirect impacts on aquatic communities. University Press of New England, Hanover, NH.

Mills, E.L., J.H. Leach, J.T. Carlton, and C.L. Secor. 1993. Exotic species in the Great Lakes: a history of biotic crises and anthropogenic introductions. Journal of Great Lakes Research 19(1):1-54.

Mitchell, R.S. 1986. A checklist of New York State plants. Bulletin of the New York State Museum 458.

Mitchill, S.L. 1815. The fishes of New York. Transactions, Literary Philosophical Society of New York 1:355-492.

Mitchill, S.L. 1817. Report on the ichthyology of the Wallkill. American Monthly Magazine and Critical Review 1:289-290.

Mooney, H.A., and J.A. Drake (editors). 1986. Ecology of biological invasions of North America and Hawaii. Ecological Studies 58.

Morgan, D.E., M. Kesser, J.F. Foertch, and J.M. Vozanik. 1991. The Asiatic clam, *Corbicula fluminea*, extends its North American distribution to the Connecticut River. Environmental Management 16(4):523.

Muenscher, W.C. 1933. Aquatic vegetation of the Upper Hudson watershed. Pp. 216-238 in A biological survey of the Upper Hudson watershed. New York State Conservation Department, Supplemental to Twenty-second Annual Report (1933).

Muenscher, W.C. 1935a. Aquatic vegetation of the Mohawk watershed. Pp. 228-249 in A biological survey of the Mohawk-Hudson watershed. New York State Conservation Department, Supplemental to Twenty-fourth Annual Report (1935).

Muenscher, W.C. 1935b. Annual report to the legislature for the year 1934. New York State Conservation Department Annual Report 24:240.

Muenscher, W.C. 1944. Aquatic Plants of the United States. Comstock Publishing Company, Inc., Ithaca, NY. 3

Myers, G.S. 1925. Introduction of the European bitterling (*Rhodeus*) in New York and of the rudd (*Scardinius*) in New Jersey. Copeia 141:20-21.

Nalepa, T.F., and D.W. Schloesser (editors). 1993. Zebra mussels: biology, impacts, and control. CRC Press, Inc., Boca Raton, FL.

Nevin, F.R., and H.K. Townes. 1935. Studies of invertebrate forage organisms in selected areas with notes on the effect of pollution on them. Pp. 214-227 in A biological survey of the Mohawk-Hudson watershed. New York State Conservation Department, Supplemental to Twenty-fourth Annual Report (1934).

Nuttall, T. 1818. The genera of North American plants. [Facsimile of the 1818 edition]. Classica Botanica Americana 7. 1971. J. Ewan (ed.). Hafner Publishing Company, New York.

Office of Technology Assessment. 1993. Harmful non-indigenous species in the United States. Summary Report to the United States Congress.

Olsen, T.M., D.M. Lodge, G.M. Capelli, and R.J. Houlihan. 1991. Mechanisms of impact of an introduced crayfish (*Orconectes rusticus*) on littoral congeners, snails, and macrophytes. Canadian Journal of Fisheries and Aquatic Sciences 48:1853-1861.

Ortmann, A.E. 1906. The crayfishes of the state of Pennsylvania. Memoirs of the Carnegie Museum (Pittsburgh) 3(2):387-406.

Page, L.M., and B.M. Burr. 1991. Peterson field guide: freshwater fishes. Houghton Mifflin Company, Boston, MA.

Paige, E.W. 1864. Catalogue of the flowering plants of Schenectady County. Van Benthuysen's Steam Printing House, Albany, NY.

Peckarsky, B.L., P.R. Fraissinet, M.A. Penton, and D.J. Conklin, Jr. 1990. Freshwater macroinvertebrates of northeastern North America. Cornell University Press, Ithaca, NY.

Pennak, R.W. 1978. Fresh-water invertebrates of the United States. 2nd edition. John Wiley & Sons, Inc., New York.

Pickett, J.F., and R.J. Sloan. 1985. *Procambarus (Ortmannicus) acutus acutus* Girard (Decapoda: Cambaridae) introduced into northeastern New York. Northeastern Environmental Science 4:39-41.

Pieterse, A.H., and K.J. Murphy. 1990. Aquatic weeds: The ecology and management of nuisance aquatic vegetation. Oxford University Press, London.

Powell, J., and D.L. Strayer. 1992. Zebra mussels in the Hudson. Tidal Exchange 2(3):1-2.

Rawinski, T.J. 1982. The ecology and management of purple loosestrife (*Lythrum salicaria* L.) in central New York. Master of Science thesis, Cornell University, Ithaca, NY. 88 pages.

Reed, C.F. 1970. Selected weeds of the United States. United States Department of Agriculture, Agriculture Research Service, Agriculture Handbook Number 366.

Reed, C.F. 1977. History and distribution of Eurasian water-milfoil in United States and Canada. Phytologia 36(5):417-436.

Rehder, H.A. 1937. *Congeria leucophaeata* (Conrad) in the Hudson River. Nautilus 50(4):143.

Reider, R.H. 1979. Occurrence of the silver lamprey in the Hudson River. New York Fish and Game Journal 26(1):93.

Ristich, S.S., M. Crandall, and J. Fortier. 1977. Benthic and epibenthic macroinvertebrates of the Hudson River: distribution, natural history, and community structure. Estuarine and Coastal Shelf Science 5:255-266.

Robertson, I.C.S., and C.L. Blakeslee. 1948. The Mollusca of the Niagara frontier region. Bulletin of the Buffalo Society of Natural Sciences 19(3):1-191.

Robins, C.R., R.M. Bailey, C.E. Bond, J.R. Brooker, E.A. Lachner, R.N. Lea, and W.B. Scott. 1991. Common and scientific names of fishes from the United States and Canada. American Fisheries Society Special Publication 20.

Scales, P., and A. Bryan. 1979. Transport of *Myriophyllum spicatum* fragments by boaters and assessment of the 1978 Boat Quarantine Program. Studies on aquatic macrophytes 27. Water Investigations Branch, Ministry of Environment, British Columbia, Canada.

Schmidt, R.E., J.M. Samaritan, and A. Pappantoniou. 1981. Status of the bitterling, *Rhodeus sericeus*, in Southeastern New York. Copeia 1981(2):481-482.

Seymour, H., S. Green, and R.B. Roosevelt. 1869. Report of the Commissioners of Fisheries of the state of New York. Weed, Parsons, and Company, Albany, NY.

Simpson, K.W. 1976. A water quality evaluation of the Hudson River, based on the collection and analysis of macroinvertebrate communities. 47 pp. in Hudson River ecology: fourth symposium on Hudson River ecology, held at Bear Mountain, New York, March 28-30, 1976, sponsored by the Hudson River Environmental Society, Inc.

Simpson, K.W., and L.E Abele. 1984. *Ripistes parasita* (Schmidt) (Oligochaeta: Naididae), a distinctive oligochaete new to North America. Freshwater Invertebrate Biology 3(1):36-41.

Simpson, K.W., R.W. Bode, J.P. Fagnani, and D.M. De Nicola. 1984. The freshwater macrobenthos of the main channel, Hudson River. Part B. Biology, taxonomy and distribution of resident macrobenthic species. Final Report to Hudson River Foundation, Grant 8/83A/39.

Simpson, K.W., J.P. Fagnani, R.W. Bode, D.M. De Nicola, and L.E. Abele. 1986. Organism-substrate relationships in the main channel of the lower Hudson River. Journal of North American Benthological Society 5:41-57.

Smith, C.L. 1976. The Hudson River fish fauna. 11 pp. in Hudson River ecology: fourth symposium on Hudson River ecology, held at Bear Mountain, New York, March 28-30, 1976, sponsored by the Hudson River Environmental Society, Inc.

Smith, C.L. 1985. The inland fishes of New York State. The New York State Department of Environmental Conservation, New York.

Smith, C.L., and T.R. Lake. 1990. Documentation of the Hudson River fish fauna. American Museum Novitates. Number 2981.

Smith, D.G. 1979. New locality records of crayfish from the middle Hudson River system. Ohio Journal of Science 79:133-135.

Smith, D.G. 1982. The zoogeography of the freshwater mussels of the Taconic and southern Green Mountain region of northeastern North America (Mollusca: Pelecypoda: Unionacea). Canadian Journal of Zoology 60:261-267.

Smith, D.G. 1983. Notes on Mississippi River basin Mollusca presently occurring in the Hudson River system. Nautilus 97:128-131.

Smith, D.G. 1985. A study of the distribution of freshwater mussels (Mollusca: Pelecypoda: Unionide) of the Lake Champlain drainage in northwestern New England. American Midland Naturalist 114:19-29.

Smith D.G. 1991. Keys to the freshwater macroinvertebrates of Massachusetts. University of Massachusetts, Amherst, MA.

Smith, S.J. 1965. Checklist of the grasses of New York State. New York State Museum Bulletin 403.

Snelson, F.F. 1968. Systematics of the cyprinid fish *Notropis amoenus*, with comments on the subgenus *Notropis*. Copeia 1968(4):776-802.

Spidle, A.P., J.E. Marsden, and B. May. 1994. Identification of the Great Lakes quagga mussel as *Dreissena bugensis* from the Dneiper River, Ukraine, on the basis of allozyme variation. Canadian Journal of Fisheries and Aquatic Sciences 51:1485-1489.

Strayer, D. 1987. Ecology and zoogeography of the freshwater mollusks of the Hudson River basin. Malacological Review 20:1-68.

Strayer, D. 1990. Aquatic Oligochaeta. Pp. 373-397 in B.L. Peckarsky, P.R. Fraissinet, M. A. Penton, and D.J. Conklin, Jr. (eds.). Freshwater macroinvertebrates of the northeastern North America. Cornell University Press, Ithaca, NY.

Strayer, D., and J. Powell. 1992. Appearance and spread of the zebra mussel in the Hudson River estuary in 1991. *Dreissena polymorpha* Information Review 3(2):1, 3.

Stuckey, R.L. 1970. Distributional history of *Epilobium hirsutum* (Great Hairy Willow-herb) in North America. Rhodora 72:164-181.

Stuckey, R.L. 1979. Distributional history of *Potamogeton crispus* (curly pondweed) in North America. Bartonia 46:22-42.

Stuckey, R.L. 1980. Distributional history of *Lythrum salicaria* (purple loosestrife) in North America. Bartonia 46:3-20.

Stuckey, R.L., and W.L. Phillips. 1970. Distributional history of *Lycopus europaeus* (European water horehound) in North America. Rhodora 72(791):351-369.

Swingle, H. 1957. Control of pond weeds by use of herbivorous fishes. Proceedings, Annual Meeting of the Southern Weed Conference 10:11-17.

Taylor, W.R. 1969. A revision of the catfish genus *Noturus* Rafinesque with an analysis of higher groups in the Ictaluridae. Bulletin of the United States National Museum 282.

Thompson, J.H. (editor). 1966. Geography of New York State. Syracuse University Press, Syracuse, NY.

Torrey, J. 1843. A flora of the State of New York, comprising full descriptions of all the indigenous and naturalized plants hitherto discovered in the state; with remarks on their economical and medical properties. Volume I. D. Appleton and Company, New York.

Townes, H.K. 1937. Studies on the food organisms of fish. Pp. 217-230 in A biological survey of the Lower Hudson watershed. New York State Conservation Department, Supplemental to 26th Annual Report.

Trautman, M.B. 1981. The fishes of Ohio with illustrated keys. Revised edition. The Ohio State University Press, Columbus, OH.

Tucker, G. 1994. Two galloping grasses. New York Flora Association Newsletter 5(4):2-3.

Turgeon, D.D. (editor). 1988. Common and scientific names of aquatic invertebrates from the United States and Canada: mollusks. American Fisheries Society Special Publication 16.

Usher, G. 1974. A dictionary of plants used by man. Constable and Company, London.

Verill, A.E. 1874. Synopsis of North American freshwater leeches. Pp. 666-689 in U.S. Fisheries Commission Report for 1872-73, Part 2.

Voss, E.G. 1985. Michigan flora. Part II: dicots (Saururaceae-Cornaceae). Cranbrook Institute of Science, Bulletin 59, Bloomfield Hills, MI.

Wales, J.H. 1939. General report of investigations on the McCloud River drainage in 1938. California Fish and Game Journal 25:272-309.

Walton, W.C. 1993. Invasion of the Hudson River estuary (NY, U.S.A.) by the zebra mussel, *Dreissena polymorpha*. Master of Science thesis, Rutgers, The State University of New Jersey, NJ. 118 pp.

Wibbe, H. 1886. Notes from Schenectady, New York. Bulletin of the Torrey Botanical Club 13:79.

Winnell, M.H., and D.J. Jude. 1987. Benthic community structure and composition among rocky habitats in the Great Lakes and Keuka Lake, New York. Journal of Great Lakes Research 13(1):3-17.

Wood, W.M. 1892. *Paludina japonica* Mart. for sale in the San Francisco Chinese markets. Nautilus 5:114-115.